Isotopic and elemental tracers of Cenozoic climate change

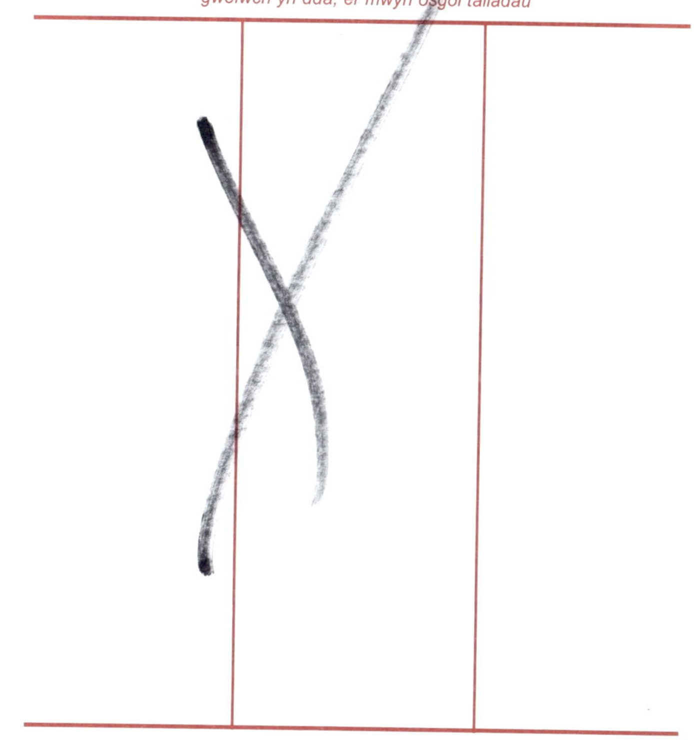

THE GEOLOGICAL SOCIETY OF AMERICA

Special Paper 395

3300 Penrose Place, P.O. Box 9140 ▪ Boulder, Colorado 80301-9140, USA

2005

Published by The Geological Society of America, Inc.
3300 Penrose Place, P.O. Box 9140, Boulder, Colorado 80301-9140, USA
www.geosociety.org

Printed in U.S.A.

GSA Books Science Editor: Abhijit Basu

Library of Congress Cataloging-in-Publication Data

Isotopic and elemental tracers of Cenozoic climate change / edited by Germán Mora, Donna M. Surge..
p. cm. -- (Special paper ; 395)
Includes bibliographical references.
ISBN 0-8137-2395-7 (pbk.)
1. Paleoclimatology--Cenozoic. 2. Climatic changes. 3. Isotope geology. I. Mora, Germán, 1970- II Surge, Donna M. III. Special papers (Geological Society of America) ; 395.

QC884.2.C5I86 2005
551.6'09'012--dc22

20050501565

Cover: Map showing the locations of the sites discussed in this volume and some representative diagrams and figures from the studies included in this volume

10 9 8 7 6 5 4 3 2 1

Contents

Preface

This volume resulted from a session entitled "Isotopic and Elemental Tracers of Cenozoic Climate Change," which was held at the 2001 Annual Meeting of the Geological Society of America. The aim of the session was to provide a forum to exchange ideas and stimulate discussion of techniques, opportunities, and new developments in geochemical approaches to reconstructing paleoclimate. Although geochemical techniques have been employed in the assessments of past climatic changes for a long time, it was apparent to us from the discussions at the meeting that geochemistry keeps pushing the envelope in the interpretation and resolution of paleoclimatological reconstructions. The papers included in this volume are excellent examples of this issue, because they explore theoretical and practical frameworks to analyze and interpret isotopic and elemental signals across a range of substrates. These papers also provide new avenues to improve paleoclimate proxy records in fields such as oceanography, limnology, hydrology, aquatic ecology, and pedology.

A number of papers in this volume focus on marine systems. D. Thomas explores, for instance, the use of neodymium isotope composition of fossil teeth as a proxy to reconstruct ancient deep-water mass compositions. F. Andrus et al. analyze the radiocarbon content of mollusks collected from offshore Peru to evaluate El Niño/southern oscillation–related changes in upwelling intensity along this region. B. Schöne and collaborators introduce a new graphical model to study bivalve mollusk shells and perform high-resolution reconstructions of temperature, salinity, and nutrient content.

Some papers in the volume focus on terrestrial settings. G. Mora and L. Hinnov, for instance, explore the possibility of using sulfur speciation in lake sediments as a proxy for precipitation rates in tropical regions. F. Serefiddin and collaborators develop a technique to determine the hydrogen isotope composition of fluid inclusions trapped in speleothems to evaluate changes in the isotopic composition of rainwater. N.J. Tabor and C.J. Yapp analyze the oxygen, carbon, and hydrogen isotope composition of karst-fill calcite and goethite to evaluate whether these two minerals can provide environmental conditions of the time intervals when they formed.

This volume is by no means a comprehensive manual of methods and techniques for paleoclimatic interpretations, but rather it serves to provide illustrative examples of the dynamism in which the field of geochemistry is contributing to understand past changes in climate. We appreciate the contributions of the authors and the thorough comments provided by reviewers that were instrumental in ensuring the quality of the manuscripts presented in this volume. We also valued the support provided by Iowa State University and The University of North Carolina at Chapel Hill. Finally, we would like to thank Abhijit Basu and the editorial staff of the Geological Society of America for helping us in bringing this volume to completion.

Germán Mora and Donna Surge

Geological Society of America
Special Paper 395
2005

Reconstructing ancient deep-sea circulation patterns using the Nd isotopic composition of fossil fish debris

Deborah J. Thomas*
Department of Geological Sciences, University of North Carolina, Chapel Hill, North Carolina 27599-3315, USA

ABSTRACT

Understanding the role of thermohaline circulation in past climate depends on proxy-based reconstructions of deep-water mass composition. A growing body of data indicates that the neodymium (Nd) isotopic composition of fossil fish debris found in deep-sea sediments can be used as a gauge of deep-water mass composition. This paper reviews the basis for the use of fossil fish debris Nd isotopic values as a proxy for ancient deep-water mass composition and then presents an example of how ancient thermohaline circulation patterns are reconstructed using records of fish debris Nd from deep-sea sediment cores.

Fish teeth and debris acquire enhanced Nd concentrations (~100s to ~1000 ppm) during an early diagenetic reaction at the sediment/water interface. Data published in 2004 by Martin and Scher and presented in this work confirm the assertion that fish teeth and debris record and retain a deep-water mass signal. The widespread stratigraphic and geographic occurrence of fish remains in deep-sea sediments enables construction of much higher-resolution records than previously afforded by Nd isotopic records derived from Fe-Mn crusts. The utility of fish debris Nd isotopic records is apparent when applied to the late Paleocene–early Eocene interval ca. 55 Ma, an interval of time poorly represented in Fe-Mn crust records. Fish debris records from a suite of deep-sea sedimentary sections indicate a mode of thermohaline circulation fundamentally different from the modern–deep waters formed primarily in the Southern Ocean, with no source of deep-water formation in the North Atlantic. This work reinforces the validity and effectiveness of fish debris Nd isotopic records as a tool for reconstructing ancient thermohaline circulation patterns.

Keywords: Nd isotopes, paleoceanography, fish teeth geochemistry, Ocean Drilling Program.

INTRODUCTION

A large proportion of net heat transport from the equator to the poles is achieved through the ocean's thermohaline circulation (e.g., Broecker, 1997). Thus thermohaline circulation plays an important role in maintaining the equator-to-pole thermal gradient and is a significant factor in the internal modulation of global climate. In order to understand the dynamics of past climates, particularly the warm, "no-analog" greenhouse climates of the mid-Cretaceous and early Paleogene, we must be able to characterize the mode of thermohaline circulation during those intervals.

*Current address: Department of Oceanography, Texas A&M University, College Station, Texas 77843-3146, USA.

Thomas, D.J., 2005, Reconstructing ancient deep-sea circulation patterns using the Nd isotopic composition of fossil fish debris, *in* Mora, G., and Surge, D., eds., Isotopic and elemental tracers of Cenozoic climate change: Geological Society of America Special Paper 395, p. 1–11, doi: 10.1130/2005.2395(01). For permission to copy, contact editing@geosociety.org.

Modern deep-water masses can be characterized and tracked by any one of a variety of conservative or nonconservative properties, such as potential temperature, density, dissolved oxygen, or dissolved nutrients (e.g., Schmitz, 1995, and references therein). However, the composition and pathway of ancient water masses must be reconstructed from the geologic record via sedimentary and geochemical proxies. Much use has been made of the oxygen and carbon isotopic composition of calcite tests secreted by bottom-dwelling foraminifera to discern paleotemperature and compositional information of ancient deep waters. However, these proxies are susceptible to alteration by calcite diagenesis, and in many deep-sea environments these proxies are unretrievable from the sedimentary record due to carbonate dissolution (a common problem for abyssal locations that lie below the calcite compensation depth and hence preserve no calcite microfossils). In addition, the signal of deep-water mass aging contained in $\delta^{13}C$ values can be overprinted by surface water productivity. Also, deep-water $\delta^{18}O$ values record a combination of temperature and salinity conditions in a given water mass, and water masses may have similar $\delta^{18}O$ compositions despite differing temperatures and salinities.

Thus an alternate proxy for ancient deep-water mass composition is needed. Such a proxy should have a well-characterized marine geochemical behavior as well as a widespread geological record and geographic distribution. Ideally this proxy should be resistant to burial diagenesis, maintaining the original paleoenvironmental signal, and should be relatively easy to analyze with high precision. The focus of this contribution is on the utility and application of the rare earth element (REE) neodymium (Nd) recorded by biogenic apatite (fish teeth and bones) as a proxy for deep-water mass composition.

Marine Geochemical Cycling of Nd

Neodymium is a light rare earth element with seven naturally occurring isotopes. Geologists are primarily interested in the ratio of the radiogenic isotope ^{143}Nd to the stable isotope ^{144}Nd, the former produced by the decay of ^{147}Sm with a half-life of 1.06×10^{11} years. The evolution of $^{143}Nd/^{144}Nd$ in a given lithology results from different initial Sm and Nd concentrations acquired during mineral formation. Because of the relatively long half-life of ^{147}Sm, geological variations in $^{143}Nd/^{144}Nd$ values are small, typically in the 4th, 5th, and 6th decimal place of 0.512.

To facilitate interpretation and presentation of Nd isotopic data, we employ the epsilon notation, ε_{Nd}, which normalizes the $^{143}Nd/^{144}Nd$ value of a geologic sample to that of the bulk earth where:

$$\varepsilon_{Nd} = ((^{143}Nd/^{144}Nd_{sample} / 0.512638) \times 10{,}000) - 1$$

(DePaolo and Wasserburg, 1976). Old, continental rocks that contain relatively low concentrations of Sm are characterized by low $^{143}Nd/^{144}Nd$ values and hence very negative, nonradiogenic ε_{Nd} values. Younger, mantle-derived rocks such as arc terranes and mid-ocean ridge basalts have higher initial Sm concentrations and consequently higher $^{143}Nd/^{144}Nd$ values and radiogenic ε_{Nd} values.

Neodymium is supplied to the oceans through weathering and drainage of subaereally exposed rocks (e.g., Goldstein and Jacobsen, 1988; Elderfield and Greaves, 1982; Halliday et al., 1992; Jones et al., 1994). The distribution of Nd isotopic values in deep-ocean waters (Fig. 1) demonstrates distinct interbasinal differences. The most negative, nonradiogenic ε_{Nd} values occur in the North Atlantic (~–12 to –14) imparted by weathering and drainage of the ancient Canadian rocks. In contrast, the most radiogenic ε_{Nd} values exist in the North Pacific (~–5), a consequence of the weathering and drainage of circum-Pacific arc terranes. The Indian and Southern Oceans are characterized by ε_{Nd} values intermediate between the North Atlantic and North Pacific end members. These interbasinal differences imply that the oceanic residence time of Nd is short with respect to the mixing time of the oceans (~1500 yr; Broecker et al., 1960). This short oceanic residence time of Nd, estimated to be ~1000 yr (e.g., Tachikawa et al., 1999), renders Nd a useful tracer of deep-water mass

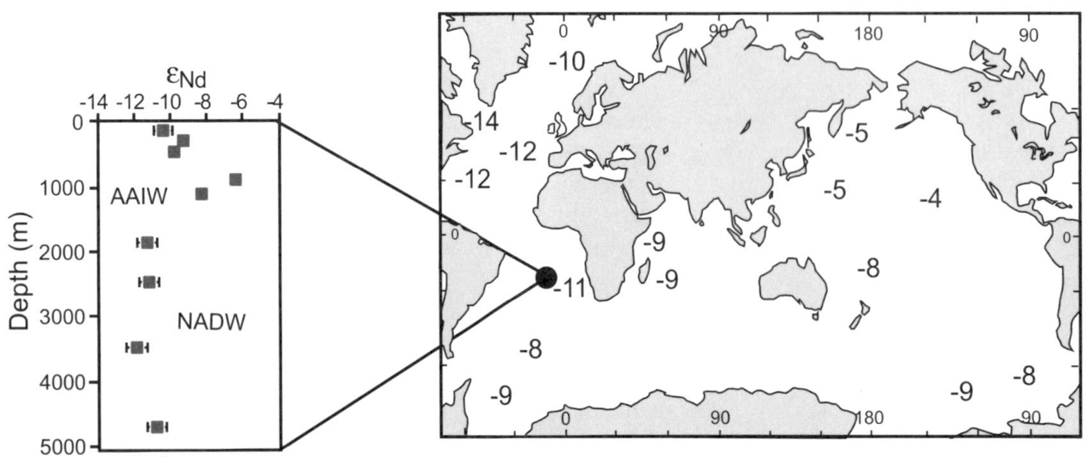

Figure 1. Map of the modern distribution of ε_{Nd} in oceanic deep waters (>3000 m), with a vertical profile of water column ε_{Nd} values. Deep-water values compiled by Jones et al. (1994), and the profile is from Jeandel (1993). NADW—North Atlantic Deep Water; AAIW—Antarctic Intermediate Water.

transit (e.g., Piepgras and Wasserburg, 1982; Bertram and Elderfield, 1993; Jeandel, 1993), similar to the temperature, salinity, and nutrient characteristics of modern water masses.

The Nd isotopic composition of individual deep-water masses is derived from the composition of dissolved, and to a lesser extent, suspended materials draining into the source regions of the water masses (Goldstein and Jacobsen, 1988; Elderfield et al., 1990; Sholkovitz, 1993). For example, North Atlantic Deep Water forms as dense waters from the Nordic Seas (ε_{Nd} ~–9) flow southward and mix with sinking waters in the Labrador Sea with a surface ε_{Nd} value as low as ~–26 (Piepgras and Wasserburg, 1987). The resulting water mass has an ε_{Nd} signature of ~–12 to –13, which can be used to track North Atlantic Deep Water throughout its deep-sea transit (Fig. 1).

Southern Ocean waters (both Antarctic Intermediate Water and Antarctic Bottom Water) have a more radiogenic signature than North Atlantic Deep Water, derived from North Atlantic Deep Water mixing with the waters flowing eastward through the Drake Passage (Piepgras and Wasserburg, 1982). Thus Antarctic waters have an ε_{Nd} value of ~–9 (Piepgras and Wasserburg, 1982).

The most radiogenic ε_{Nd} values are found in the surface waters of the North and South Pacific, which have a characteristically radiogenic signature of ~0 to –4 (Piepgras and Wasserburg, 1982; Piepgras and Jacobsen, 1988). This reflects the average fluvial input to the Pacific of –2.9 to –3.7 (Goldstein and Jacobsen, 1988). But depth profiles of Nd isotopic composition indicate considerable stratification of South Pacific waters compared to the North Pacific (Fig. 2). Analyses from South Pacific Station 80 indicate very radiogenic surface waters of ε_{Nd} ~0 underlain by much more nonradiogenic ε_{Nd} values of ~–8 at 4500 m water depth (Piepgras and Wasserburg, 1982). The more negative, nonradiogenic bottom water signature reflects the northward flow of Antarctic bottom waters into the Pacific (Piepgras and Jacobsen, 1988).

However, depth profiles of Nd isotopic composition in the North Pacific are significantly different. While there is a slight trend toward more negative values at depth, much less stratification of ε_{Nd} values is evident in North Pacific profiles (Fig. 2) despite the existence of distinct intermediate, deep, and bottom water masses (e.g., Tomczak and Godfrey, 1994). The dominance of "Pacific" ε_{Nd} values and the absence of an Antarctic Bottom Water ε_{Nd} signature (i.e., ~–8) in the bottom waters of the North Pacific is probably the consequence of very slow deep-water renewal in this region. Slow deep-water renewal in the North Pacific results from a combination of factors: (1) relatively high precipitation in the North Pacific results in sea-surface salinities too low to enable downwelling (~33 ppt as compared to ~35 ppt in the North Atlantic; Levitus, 1982); (2) the Bering Strait sill does not permit dense, cold Arctic bottom waters to enter the North Pacific (Tomczak and Godfrey, 1994); and (3) bottom waters formed in the Ross Sea (the Pacific sector of the Southern Ocean) are prevented from flowing northward into the Pacific basin by the eastward flow of the Antarctic circumpolar current and the mid-ocean ridge (Tomczak and Godfrey, 1994). Thus vertical exchange of Nd between Pacific intermediate waters and the underlying deep- and bottom-water masses is able to dominate over advective transport of the Antarctic Bottom Water ε_{Nd} signal, imparting the relatively radiogenic ε_{Nd} signature (~–5) to the deep North Pacific.

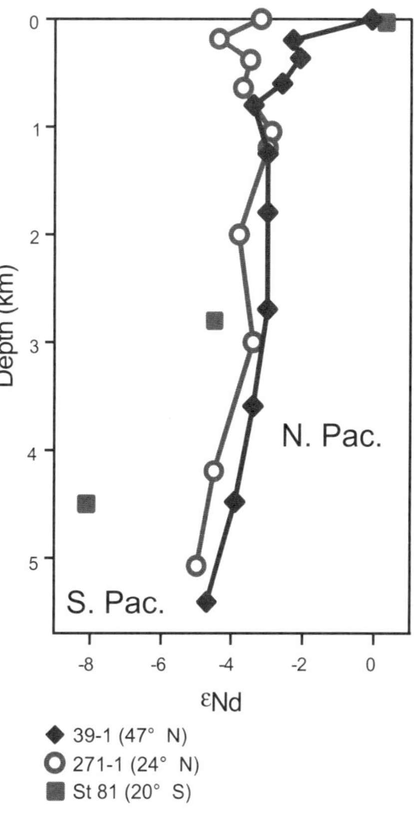

Figure 2. Vertical profiles of seawater ε_{Nd} for three Pacific stations. North Pacific data from Piepgras and Jacobsen (1988) and South Pacific data from Piepgras and Wasserburg (1982). The difference between deep-water ε_{Nd} in the North and South Pacific reflects the incursion of less radiogenic Southern Ocean waters (ε_{Nd} ~ –8) into the South Pacific.

Investigation of Ancient Deep-Water Nd—The Record from Fe-Mn Crusts

Most investigations of paleo-Nd isotopic composition employ analyses of the layers of Fe-Mn crusts (e.g., Burton et al., 1997), which precipitate directly from seawater. Such records have established the Cenozoic evolution of deep-water masses in the North Atlantic, Indian, and North Pacific basins (Burton et al., 1997; O'Nions et al., 1998; Ling et al., 1997; Fig. 3). The spatial distribution of modern deep-ocean ε_{Nd} values has persisted throughout much of the Cenozoic, with the most negative, nonradiogenic values recorded in the Atlantic (~–10 to –11), the most radiogenic ε_{Nd} values found in the Pacific (~–4 to –5), and intermediate ε_{Nd} values occurring in the Indian Ocean (~–7 to –8). The trend toward modern nonradiogenic Atlantic ε_{Nd} values (~–12 to –13) only began ca. 5 Ma as North Atlantic Deep Water produc-

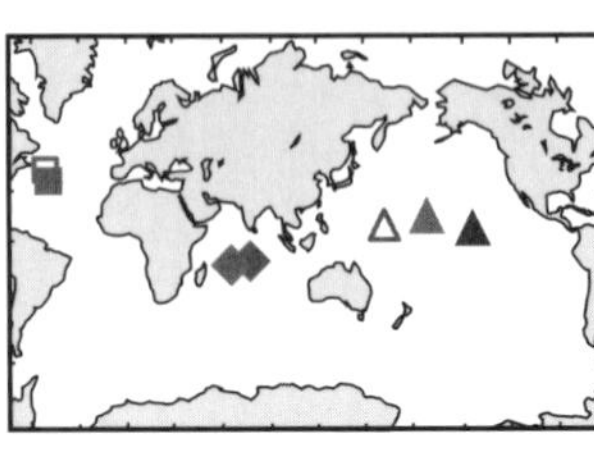

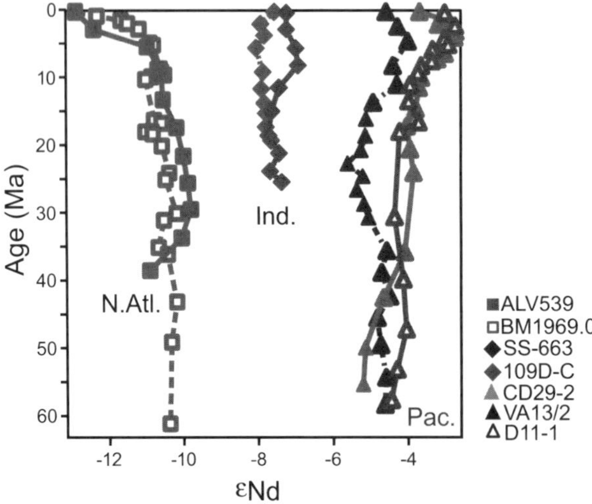

Figure 3. Cenozoic records of seawater ε_{Nd} from analyses of Fe-Mn crusts. Data from crusts ALV539 and 109D-C generated by O'Nions et al. (1998), crust BM1969.05 from Burton et al. (1997), crust SS-663 from O'Nions et al. (1998), and crusts CD29-2, VA13/2, and D11-1 from Ling et al. (1997).

tion intensified with the gradual emplacement of the Panamanian Isthmus (Burton et al., 1997).

The long-term crust records have laid the framework for reconstructing the nature of thermohaline circulation throughout the Cenozoic. However, Fe-Mn crusts precipitate very slowly, on the order of a few mm per m.y. Therefore each analysis of a crust layer represents an average of approximately a million years of averaged seawater chemistry. Changes in the mode of thermohaline circulation have been postulated to occur on as short as a millennial time scale (e.g., Broecker, 1998), rendering their resolution impossible using the long-term crust records. In addition, the analyzed crusts only provide limited geographic coverage of the ocean basins.

In this contribution, I discuss fossil fish teeth and debris as an alternative, high-resolution recorder of deep-water mass Nd.

Fossil Fish Debris—The High-Resolution Alternative

The teeth and bones of fossil fish are useful for paleo-Nd investigations because of their relatively high Nd concentrations (100–1000 ppm) (e.g., Wright et al., 1984; Shaw and Wasserburg, 1985; Staudigel et al., 1985), as well as their resistance to dissolution in corrosive bottom waters. In addition, they are present, albeit rare, in most deep-sea sedimentary sections, thus extending the geographic and stratigraphic occurrence beyond that of Fe-Mn crusts. Moreover, the age of deep-sea sedimentary sections is more precisely determined than in crusts.

Fish teeth (as well as other biogenic apatite, henceforth referred to collectively as fish debris) acquire their enhanced Nd concentrations during an early diagenetic reaction at the sediment/water interface (e.g., Staudigel et al., 1985). Thus the fossil material records the Nd isotopic composition of the overlying bottom water (e.g., Wright et al., 1984; Shaw and Wasserburg, 1985; Staudigel et al., 1985; Martin and Haley, 2000). The Nd isotopic signal in fish debris has been assumed to resist diagenetic exchange with pore water during burial, and there is ample indirect evidence to support this assumption (Martin and Haley, 2000; Martin and Scher, 2004). I address this issue further by presenting new data that support the assertion that fish debris record and retain a deep-water mass signal. Then I compare the stratigraphic resolution of fish debris records with the long-term Fe-Mn crust records and discuss the fine-scale structure of the fish debris stratigraphies. Finally, I present an example of how fish debris Nd isotopic records can be used to reconstruct ancient patterns of thermohaline circulation, focusing on the late Paleocene–early Eocene interval ca. 55 Ma.

METHODS

Analytical Techniques

Recent work using the fossil remains of fish (primarily teeth, but this work also includes bones and scales that record the same signal as teeth) has employed samples from Deep Sea Drilling Project and Ocean Drilling Program cores. Teeth and fragments of fish debris were handpicked from the >63 micron size fraction of washed (disaggregated) samples from discrete

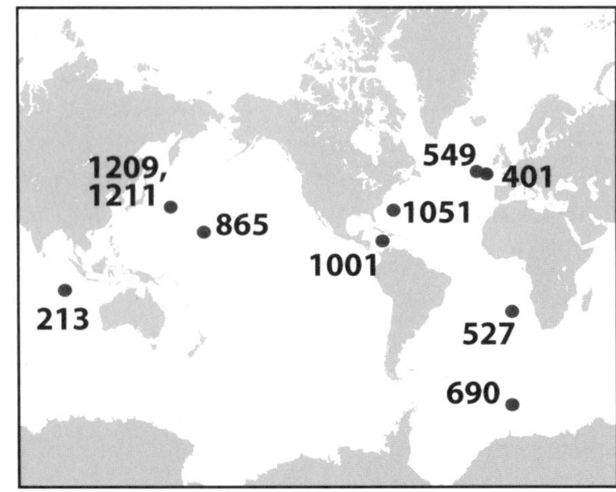

Figure 4. Map showing the present locations of the Deep Sea Drilling Project and Ocean Drilling Program Sites discussed in the text.

1–2 cm intervals of 10 Deep Sea Drilling Project and Ocean Drilling Program cores (Fig. 4). Multiple teeth/fragments were used in each analysis (in general, from 5 to 15 teeth/fragments per sample, depending on size and availability). Samples were then cleaned using an established reductive/oxidative cleaning protocol (Boyle, 1981; Boyle and Keigwin, 1985). Samples were analyzed as NdO^+ using a multicollector Micromass Sector 54 at the radiogenic isotope facility at the University of North Carolina–Chapel Hill. Monitor peak ($^{144}Nd^{16}O$) beams of ~0.5–1 V were achieved by introducing pure oxygen into the source via a leak valve. External analytical precision based upon replicate analysis of the UNC Ames Nd standard (as NdO^+) was 0.512140 ± 0.000014 (2σ) and analysis of the international standard JNd_i (Tanaka et al., 2000) yielded 0.512111 ± 0.000022, which is calibrated relative to the La Jolla standard (0.511858) as 0.512116. Reported errors are within-run 2σ values, which corresponds to a minimum uncertainty of ± 0.28 epsilon units when combined with the external precision. The procedural blank is ~50 pg and is considered negligible. Replicate analyses of eight of nine samples yielded Nd isotope values within error limits (Thomas et al., 2003).

Because fish debris record the rare earth elemental composition of seawater, it is necessary to analyze the Sm content in order to correct for any in situ production of ^{143}Nd (being the daughter of ^{147}Sm). The maximum range of $^{147}Sm/^{144}Nd$ ratios analyzed is 0.11747–0.1398 for all of the sites investigated, similar to other analyses of fish teeth (Martin and Haley, 2000). The mean $^{147}Sm/^{144}Nd$ value of 0.1286 was applied to all samples from Sites 213, 401, 527, 549, 690, 865, and 1001 to calculate $\varepsilon_{Nd}(t)$ values (Thomas et al., 2003), while a mean value of 0.132 was used to determine $\varepsilon_{Nd}(t)$ values for Site 1209 and 1211 analyses (Thomas, 2004).

The majority of the data reviewed in this contribution span the late Paleocene–early Eocene interval (ca. 51–57 Ma), however the records from Sites 1209 and 1211 extend from ca. 32 to 72 Ma. The temporal resolution of each record varies as a function of sample spacing within a given stratigraphic section. The late Paleocene–early Eocene records from Sites 213, 401, 527, 549, 690, 865, and 1001 consist of samples analyzed every 2–10 cm in the highest resolution portions of the records, corresponding to a temporal resolution of ~1–30 k.y. depending on the sedimentation rate. Lower-resolution records from Shatsky Rise Ocean Drilling Program Sites 1209 and 1211 consist of samples spaced several meters apart, thus temporal resolution is significantly lower by design (up to several m.y. between samples).

Bulk sediment samples from Deep Sea Drilling Project Site 527 (Table 1; Walvis Ridge, Southeastern Atlantic) were leached with buffered acetic acid (pH 5) to remove calcium carbonate and were subjected to a sodium citrate/sodium dithionite reductive cleaning procedure to remove the oxide coating (Jones et al., 1994) prior to Nd isotopic analysis.

Age Model

Numerical ages were assigned to deep-sea stratigraphic intervals to facilitate intersite comparison of fish debris records as well as comparison of fish debris and Fe-Mn crust records. This process involves linear interpolation of the sedimentation rate between two tie points that have been attributed numerical ages within an established chronostratigraphic framework (e.g., Berggren et al., 1995). Ages for Sites 213, 401, 527, 549, 690, 865, and 1001 were calculated based on a combination of detailed biostratigraphic and chemostratigraphic tie points established across the Paleocene/Eocene boundary interval (Thomas et al., 2003). Site 1209 and 1211 ages were calculated based on shipboard biostratigraphic analysis (Bralower et al., 2002).

RESULTS AND DISCUSSION

In this contribution I present a review of fish debris data that has been previously published in order to discuss the utility of the fish debris proxy. In addition I include previously unpublished data from the bulk sediment fraction of Deep Sea Drilling Project Site 527. To facilitate discussion of the utility and paleoceanographic applications of the fish debris Nd isotope

TABLE 1. DEEP SEA DRILLING PROGRAM SITE 527 Nd ISOTOPIC DATA FROM THE SILICATE AND FISH DEBRIS SEDIMENTARY FRACTIONS

Sample	Depth (mbsf)	Age (Ma)	Silicate				Fish Debris			
			$^{143}Nd/^{144}Nd$	%SE	ε_{Nd}	Error	$^{143}Nd/^{144}Nd$	%SE	ε_{Nd}	Error
24-1, 49-50	199.49	55.3397	0.512083	0.0009	−10.8	0.20	0.512142	0.0012	−9.7	0.20
24-1, 73-74	199.73	55.3611	0.512054	0.0007	−11.4	0.10	0.512125	0.0008	−10.0	0.16
24-2, 4-5	200.54	55.4331	0.512109	0.0007	−10.3	0.10	0.512129	0.0008	−9.9	0.16
24-2, 20-22	200.70	55.4489	0.512132	0.0007	−9.9	0.10	0.512165	0.0008	−9.2	0.16
24-2, 34-36	200.84	55.4668	0.512132	0.0007	−9.9	0.10	0.512173	0.0008	−9.1	0.16
24-2, 37-39	200.87	55.4706	0.512077	0.0006	−11.0	0.10	0.512200	0.0021	−8.6	0.40
24-2, 66-68	201.16	55.5021	0.512144	0.0011	−9.6	0.20	0.512172	0.0011	−9.1	0.22
24-3, 130-131	203.30	55.5783	0.512172	0.0007	−9.1	0.10	0.512178	0.0018	−9.0	0.36

Note: mbsf—meters below sea level; SE—standard error.

proxy of ancient deep-water composition, I have presented the results and discussion together.

Constraints on the Source of the Fish Debris Nd Isotopic Signal

An important consideration in the use of fish teeth and debris as a proxy for ancient deep- and bottom-water mass composition is whether they retain the seawater signal after burial. Toyoda and Tokonami (1990) addressed the issue of postburial uptake of REEs by fish teeth and cited the large REE concentration difference between biogenic apatite (~100 to 1000 ppm) and seawater (pg Nd per kg seawater) as evidence for continuous, diagenetic uptake of REEs in fish teeth after burial. However, recent work by Martin and Scher (2004) demonstrated that fish teeth found in older sediments do not have systematically higher Nd concentrations than those in recently deposited sediments. Such evidence argues against continuous diagenetic uptake after burial.

Additional convincing evidence that fish teeth and debris record a seawater signal comes from comparison of Pacific Ocean Drilling Program Site 1209 and 1211 fish debris data (Thomas, 2004) with Pacific crust data (Ling et al., 1997; Fig. 5). The North Pacific is the ideal setting to establish this comparison, because Nd isotopic analyses of the silicate sedimentary fraction indicate nonradiogenic values distinct from radiogenic Pacific seawater (Jones et al., 1994, 2000; Pettke et al., 2002; Fig. 5).

Fish debris analyses from Ocean Drilling Program Site 1209 (2300 m paleo– and present water depth) range from –5.3 to –2.9 epsilon units, with a mean ε_{Nd} value of –3.7. The data exhibit a trend of increasing ε_{Nd} values from –5.3 to –3.1 over the interval 279.8–204.6 m below seafloor (mbsf) (70.9–56.7 Ma). From 204.6 and 152.62 mbsf (56.7–45.8 Ma) ε_{Nd} values fluctuate between –3.4 and –2.9 then decrease to the top of the record at 114.6 mbsf (30.7 Ma).

The Nd isotopic data generated from Ocean Drilling Program Site 1211 (2900 m paleo– and present water depth) range from –5.2 to –2.8 with a mean ε_{Nd} value of –3.9. Starting at the base of the record, ε_{Nd} values increase from –3.4 to –4.4 over the interval 156.70–128.20 mbsf (69.6–58.1 Ma). This is followed upsection by an interval in which ε_{Nd} values oscillate between –3.7 and –2.8 (128.20–97.70 mbsf; 58.1–49.3 Ma). From 97.70 to 83.70 mbsf (949.3–35.8 Ma), ε_{Nd} values decrease from –5.2 to –2.9.

The similarly radiogenic ε_{Nd} values exhibited by fish debris and Fe-Mn crusts over an ~20 m.y. interval of time confirm that fish debris records the same seawater signal preserved in Fe-Mn crusts. If fish debris had acquired Nd from a source other than seawater, we would expect a more nonradiogenic signal derived from diagenetic exchange with the silicate fraction. In fact, fish debris analyses tend to exhibit more radiogenic values than the crusts when the two sets of records diverge. The differences between the Pacific crust and fish debris records (ranging up to a difference of 2 epsilon units between Crust CD29-2 and Site 1211) are most likely due to geographic differences, because the fish debris records from Sites 1209 and 1211 are further north and thus more proximal to the source of radiogenic Nd.

Analyses of fish teeth and bulk sediment Nd on a set of samples from South Atlantic Deep Sea Drilling Project Site 527

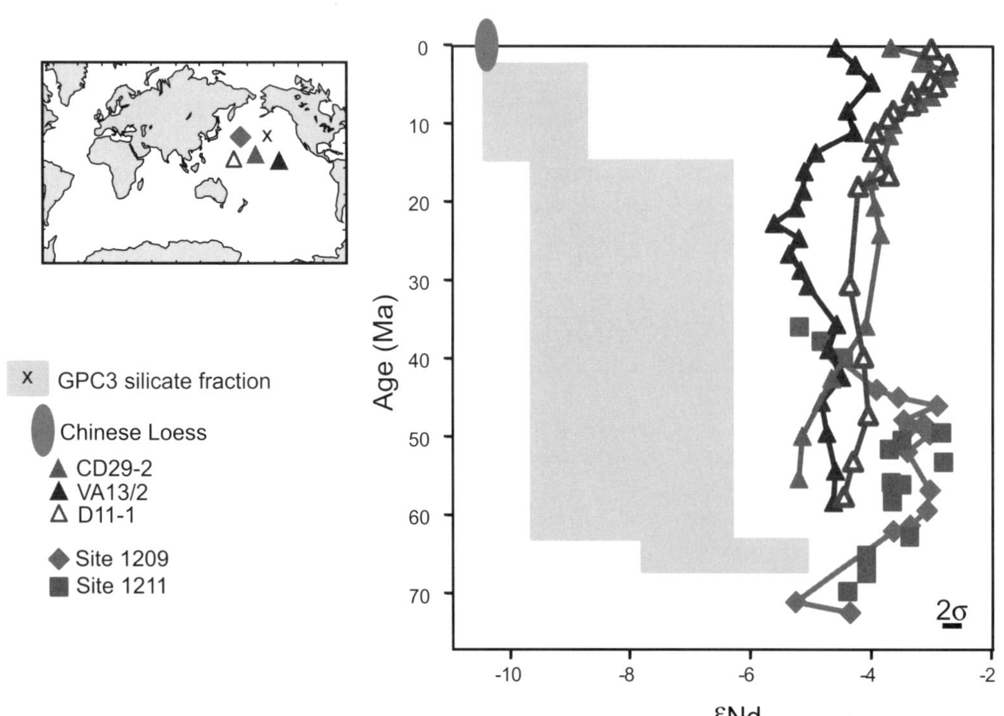

Figure 5. Comparison of fish debris data from Ocean Drilling Program Sites 1209 and 1211 with Pacific crust records (Ling et al., 1997). The shaded region represents the range of Cenozoic silicate ε_{Nd} values based on analyses of core LL44-GPC3 from Pettke et al. (2002). Core-top Chinese loess data from Jones et al. (1994). A bar representing the within-run 2σ error is included in the bottom right corner of the figure.

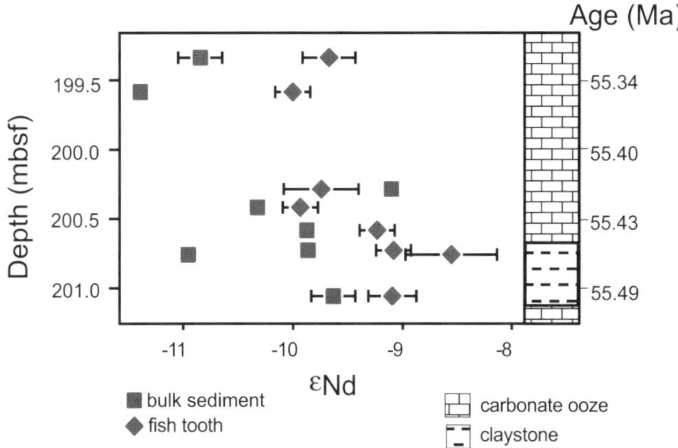

Figure 6. Suite of fish debris and bulk sediment ε_{Nd} analyses conducted on the same set of samples from Deep Sea Drilling Project Site 527. Error bars are indicated for samples in which the error bars were wider than the symbol width. Lithology is indicated at the right of the graph. mbsf—m below seafloor.

support the Pacific comparison (Fig. 6 and Table 1). The upper Paleocene–lower Eocene interval of Site 527 was chosen because it records a distinct lithologic change, from >90% carbonate to a nearly pure claystone. Both sedimentary fractions (fish debris and bulk sediment) demonstrate significant Nd isotopic variations; however there are no systematic trends that would indicate a diagenetic source of Nd for the fish debris.

The above considerations, combined with previous discussions of how fish debris acquire and retain a record of paleoseawater Nd isotopic composition (Martin and Haley, 2000; Martin and Scher, 2004), confirm the validity of fish debris as a tracer of ancient water mass composition. Fish debris record the Nd isotopic composition of the deep/bottom water mass at the sediment/water interface and retain the water mass signal after burial.

Comparison of Fish Debris Records with Crust Records

Examination of the long-term crust records (Fig. 3) gives the impression that there has been little variation in the Nd isotopic composition of oceanic deep waters during most of the past 65 m.y. This would imply that the sources of Nd to the oceans (i.e., what has been weathering and draining into deep-water formation areas) have remained constant, and it also implies that deep-water formation and circulation patterns have remained fairly constant throughout this interval. There are two reasons to doubt these scenarios. The first is tectonic: the first-order control on the evolution of thermohaline circulation patterns is the configuration of continents and ocean basins, and there has been considerable change in the shape of the ocean basins over the duration of the Cenozoic.

The second argument has to do with the nature of the crust records. As mentioned earlier, the growth of Fe-Mn crusts and nodules occurs very slowly, on the order of millimeters per m.y. Any short-term paleoceanographic changes would be smoothed out due to the slow hydrogenous uptake of Nd by the growing crust. However, Nd isotopic records derived from analyses of fish debris have the potential to resolve rapid oceanographic changes. The temporal resolution of fish debris records is limited only by stratigraphic resolution (i.e., sedimentation rate and the mixing effects of bioturbation).

Comparison of high-resolution fish debris records and crust records from the Atlantic basins (both North and South Atlantic) reveals a significant difference in the level of paleoceanographic information preserved by each (Fig. 7A). Examination of the interval ca. 51–57 Ma indicates that the fish debris record evidence for more Nd isotopic variability in the Atlantic than can be gleaned from the entire oceanwide suite of Cenozoic crust records. Thus fish debris records are much better suited for investigations of rapid or short-term paleoceanographic change.

Scales of Variability within the Nd Isotopic Records

Examination of the highest-resolution portions of the fish debris records reveals an additional component of high-frequency variation in Nd isotopic composition (Fig. 7C). Fluctuations in ε_{Nd} values of ~0.5–1.5 epsilon units are significant. However, the source of this variability is not clear. The fluctuations might have been the result of oceanographic changes (water mass composition), changes in the source of Nd to the water mass, "noise" in the system, or some combination of those three factors. While ongoing work seeks to better characterize and understand these high-frequency fluctuations, the topic still merits discussion.

The first scenario is that short-term oceanographic changes generated the high-frequency Nd isotopic changes. A possible oceanographic mechanism might be the competing influence of two end-member water mass sources on the deep-water Nd isotopic composition at each site. This scenario is feasible given that with the limitations of the age model, the high-frequency fluctuations may be correlative over broad geographic and hydrographic space (Fig. 7C). However, it is difficult to envision a mechanism that could drive basinwide changes in the composition of intermediate, deep, and bottom waters such as the periodic deep-sea ventilation proposed to explain the ~1500 yr climate cycles in glacial/interglacial records (Broecker, 1998). One difficulty with invoking the "bipolar seesaw" (Broecker, 1998) for the late Paleocene–early Eocene interval is the lack of permanent ice at both poles (e.g., Zachos et al., 1994) that might periodically alter the temperature/salinity conditions within deep-water formation regions. In addition, no significant source of deep waters existed in the North Atlantic during this time interval (Thomas et al., 2003), precluding a "bipolar" mechanism.

Alternatively, the fluctuations in the higher-resolution portions of the records might be an inherent characteristic of high-resolution deep-sea isotopic records of elements with relatively short oceanic residence times such as Nd or Pb (Christensen et al., 1997; Scher and Martin, 2001). High-frequency fluctuations in detailed deep- and bottom-water Nd isotopic records may result from short-term variations in other potential contributions

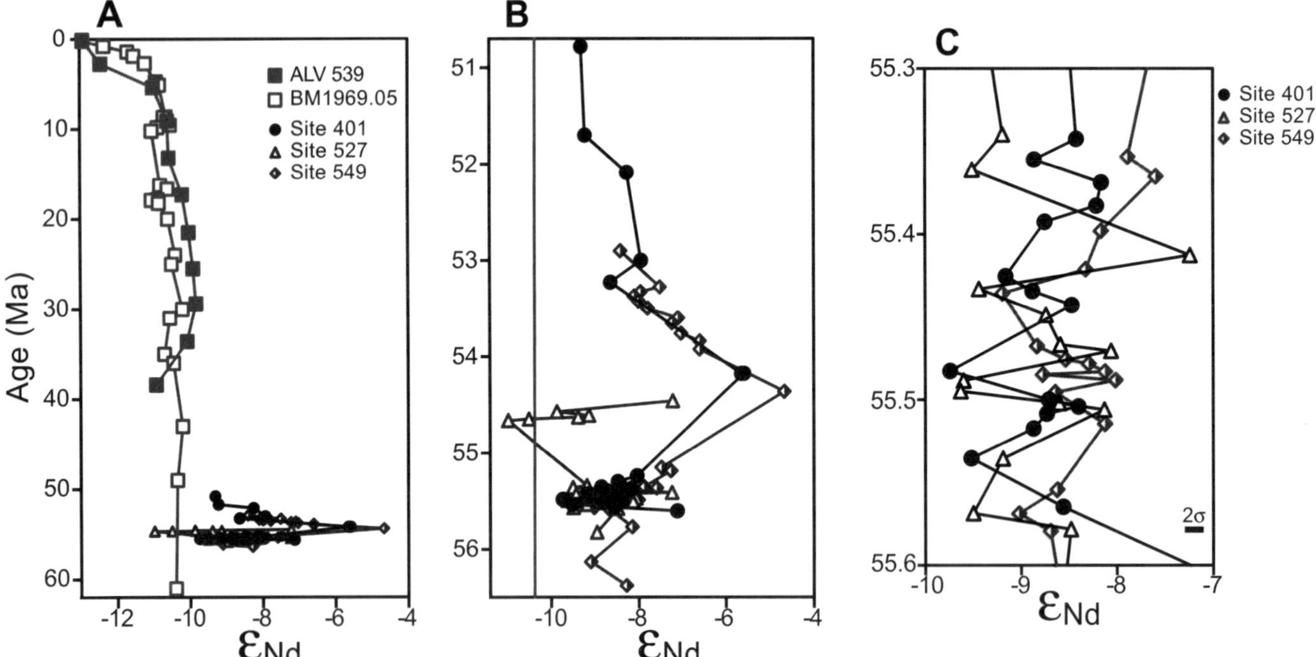

Figure 7. Graphs comparing the long-term, low-resolution North Atlantic crust data (Burton et al., 1997; O'Nions et al., 1998) with high-resolution fish debris records (Thomas et al., 2003). Panel A shows the records over the entire Cenozoic, and Panel B shows an expanded view of the interval spanned by the fish debris records. Panel C contains the high-resolution fish tooth data from Ocean Drilling Program Sites 401, 527, and 549, demonstrating rapid fluctuations in North and South Atlantic Nd isotopic composition. A bar representing the within-run 2σ error is included in the bottom right corner of panel C.

of Nd, such as mixing or entrainment of deep-water masses during transit through the ocean basins (Piepgras and Wasserburg, 1982; Bertram and Elderfield, 1993) and Nd from local fluvial inputs (e.g., Scher and Martin, 2001). These processes might only generate subtle low-amplitude Nd isotopic changes without dictating the overall deep-water signal.

Paleoceanographic Application of Fish Debris Nd Isotopic Records—An Example from the Late Paleocene–Early Eocene

The Nd isotopic composition of fossil fish debris from a given deep-sea sedimentary section provides a gauge of the deep-water mass at that location. In order to reconstruct patterns of ancient thermohaline circulation, one must first establish the Nd isotopic composition of the major ancient ocean basins for a particular stratigraphic interval (i.e., from a geographically dispersed array of deep-sea sedimentary sections for a given age). The next step is to use the ancient geographic pattern to infer the location of deep-water formation and pattern of circulation. To illustrate this strategy, I present a reconstruction of late Paleocene–early Eocene thermohaline circulation. This time interval is interesting paleoceanographically for several reasons. The tectonic and boundary conditions were much different than the later part of the Cenozoic (Fig. 8A), with an open Tethys, a closed Drake Passage, and major volcanism in the North Atlantic, which had not yet formed the Norwegian or Greenland Seas (e.g., Saunders et al., 1997). In addition, the late Paleocene–early Eocene was a time of significant global warmth and comprised part of the early Cenozoic "greenhouse" climate. Equator to pole thermal gradients were only half that of the modern sea surface temperature gradient (Zachos et al., 1994). In addition, this span of time is poorly represented in Fe-Mn crust records (Fig. 7A). Fish debris Nd isotopic analyses from upper Paleocene–lower Eocene sediments could enable a better understanding of the effects of changing oceanic gateways on the mode of thermohaline circulation as well as the role of thermohaline circulation during warm climatic intervals.

The array of deep-sea sites selected for the upper Paleocene–lower Eocene stratigraphic interval is shown in Figure 8A. Nd isotopic stratigraphies were generated for each site (Fig. 8B). Three deep-water Nd isotopic provinces are evident from the data: an "Atlantic" province, the North Pacific, and the Caribbean. Sites in the Atlantic, Indian, and Southern Oceans (Sites 213, 401, 527, 549, and 690) are generally characterized by relatively nonradiogenic Nd isotopic values (~−8.7 ε_{Nd}), thus these are grouped together as the "Atlantic" province. The occurrence of more radiogenic ε_{Nd} values of ~−4.3 at intermediate depths in the Pacific (~1500 m) suggests that this water mass had a different source of Nd than the Atlantic, Indian, and Southern

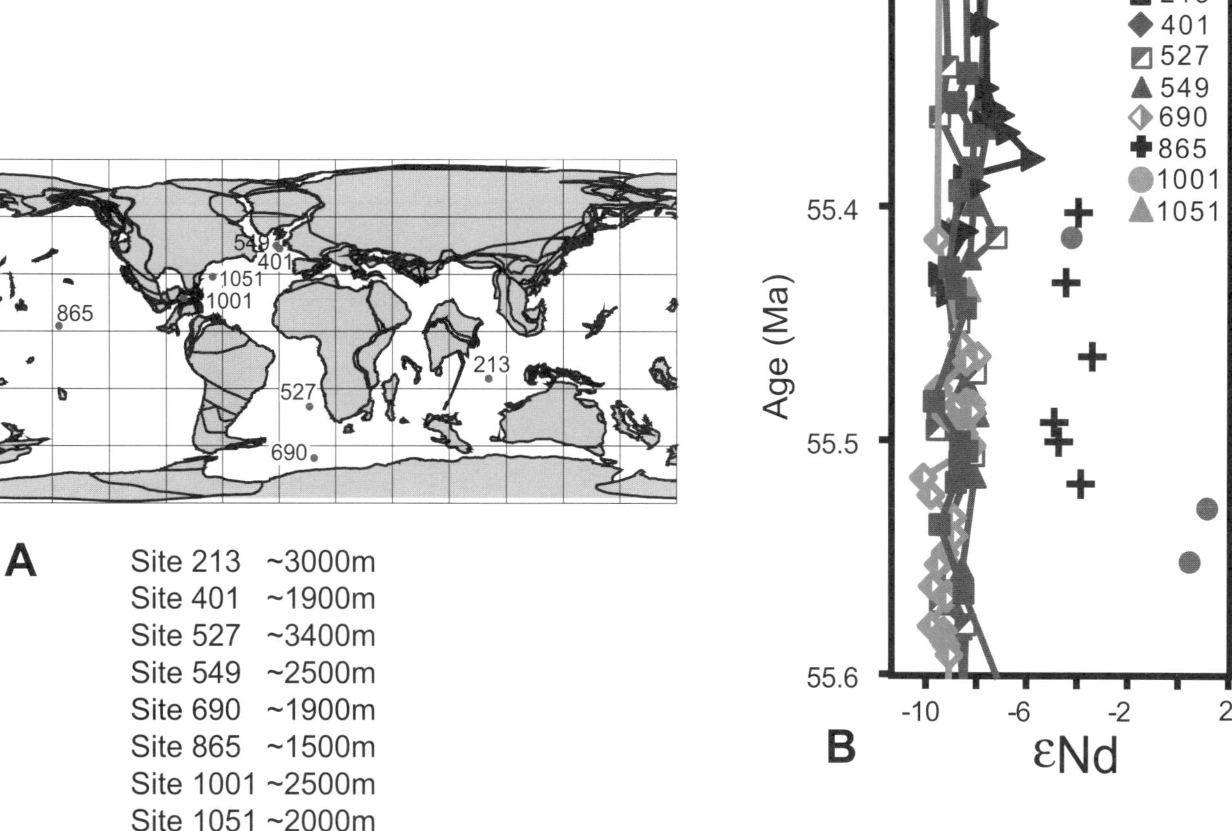

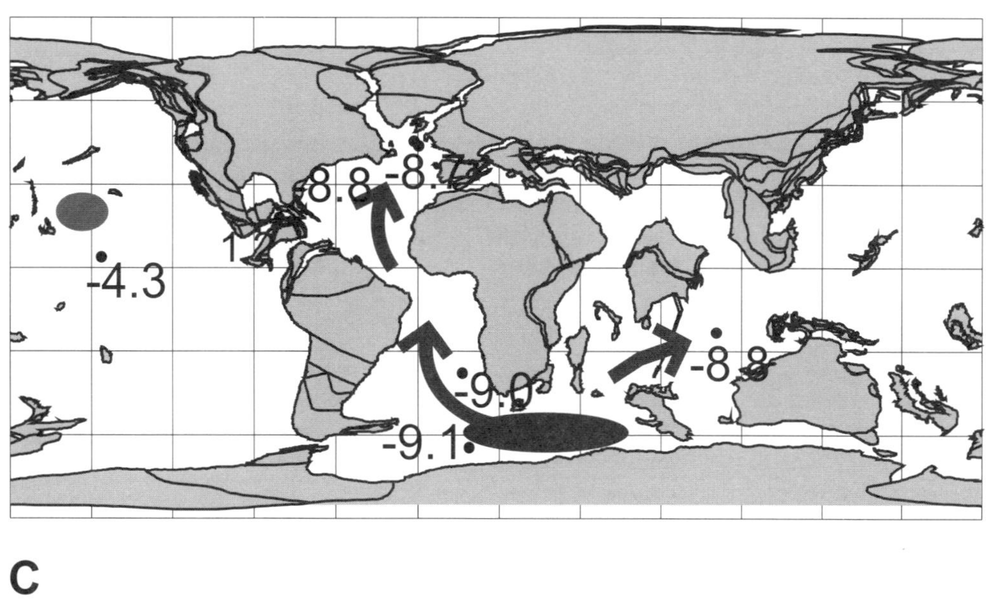

Figure 8. Reconstruction of early Paleogene (ca. 55 Ma) thermohaline circulation patterns using fish debris Nd isotopic records. (A) Paleogeographic reconstruction of the 55 Ma time slice from the Ocean Drilling Stratigraphic Network (www.odsn.de) showing the locations and paleowater depths of the Deep Sea Drilling Project and Ocean Drilling Program sites used in the investigation. (B) Fish debris records from Sites 213, 401, 527, 549, 690, 865, 1001, and 1051 spanning the Paleocene-Eocene transition. (C) Paleogeographic reconstruction showing the pattern of thermohaline circulation inferred from the Nd isotopic records.

Oceans. Caribbean Sea ε_{Nd} values averaged ~+1.2, a signature that is significantly more radiogenic than the values for the North Pacific province. Such a signature was most likely imparted by the weathering and drainage of volcanics from the contemporaneously erupting proto-Antilles arc (Sigurdsson et al., 1997; Bralower et al., 1997).

The most cohesive paleoceanographic interpretation of the general Nd isotopic records invokes a major source of Atlantic, Indian, and Southern Ocean deep waters in the surface waters of the Southern Ocean between the location of the future Drake Passage and Australia (Fig. 8C). The possibility of a contribution of northern Pacific Ocean intermediate waters (ε_{Nd} ~−4.3) to the deep Atlantic, Indian, and Southern Ocean basins can be ruled out given the nonradiogenic ε_{Nd} values that dominate them (ε_{Nd} ~−8.7). The North Atlantic can also be discounted as a significant source of deep waters based on the geographic distribution of Nd isotope values (Fig. 8C) as well as tectonic considerations. Seafloor spreading in the Labrador Sea probably only began between magnetic Chron 31 (Maastrichtian) and Chron 27 (early Paleocene; Saunders et al., 1997), and the Norwegian–Greenland Sea began opening during the latter stages of Chron 24 (early Eocene; e.g., Saunders et al., 1997). It is unlikely that the incipient basins were sites of volumetrically significant deep-water mass formation in the late Paleocene–early Eocene. Downwelling within the Indian or South Atlantic sectors of the Southern Ocean could supply deep waters to the Indian Ocean as well as the Atlantic Ocean, imparting the same Nd isotopic signature to those water masses (Fig. 8C). Although global climate was considerably warmer during the early Cenozoic and permanent Antarctic glacial ice did not exist (e.g., Zachos et al., 2001), existing oxygen isotope data indicate that Southern Ocean sea surface temperatures were still the coolest and likely densest of the oceans at that time. Thus these waters were most susceptible to downwelling, despite the overall warmer conditions.

This data set represents the most direct evidence that the deep ocean circulated in a fundamentally different manner during the early Cenozoic than during the latter part of the Cenozoic. In addition, the late Paleocene–early Eocene reconstruction reaffirms the potential applicability of fish debris for other intervals of time represented in deep-sea sediments.

CONCLUSIONS

The Nd isotopic composition of fish teeth and debris from deep-sea sediments faithfully records the Nd isotopic composition of oceanic deep waters. The seawater signal is shown to resist subsequent diagenesis. Comparison of the record of paleoseawater Nd derived from fish debris to that derived from Fe-Mn crusts reveals considerably more variability in ancient water mass composition than previously recognized. The greater temporal and geographic resolution afforded by the fish debris records is evident in a reconstruction of thermohaline circulation during the late Paleocene–early Eocene interval ca. 55 Ma. The pattern of thermohaline circulation established for the late Paleocene–early Eocene interval indicates a dominant source of deep waters forming in the Southern Ocean without a North Atlantic counterpart—a mode of circulation fundamentally different from the modern. This work reinforces the validity and potency of fish debris Nd isotopic records as a tool for reconstructing ancient thermohaline circulation patterns.

ACKNOWLEDGMENTS

I wish to thank the editors for their work on this volume, and I thank Jamie Gleason and an anonymous reviewer for their constructive reviews that improved the manuscript.

REFERENCES CITED

Berggren, W.A., Kent, D.V., Swisher, C.C., III, and Aubry, M.-P., 1995, A revised Cenozoic geochronology and chronostratigraphy, *in* Berggren, W.A., Kent, D.V., Aubry, M.-P., and Hardenbol, J., eds., Geochronology, Time Scales and Global Stratigraphic Correlations: Framework for an Historical Geology, SEPM Special Publication 54, p. 129–212.

Bertram, C.J., and Elderfield, H., 1993, The geochemical balance of the rare earth elements and neodymium isotopes in the oceans: Geochimica et Cosmochimica Acta, v. 57, p. 1957–1986, doi: 10.1016/0016-7037(93)90087-D.

Boyle, E.A., 1981, Cadmium, zinc, copper, and barium in foraminifera tests: Earth and Planetary Science Letters, v. 53, p. 11–35, doi: 10.1016/0012-821X(81)90022-4.

Boyle, E.A., and Keigwin, L.D., 1985, Comparison of Atlantic and Pacific paleochemical records for the last 250,000 years: Changes in deep ocean circulation and chemical inventories: Earth and Planetary Science Letters, v. 76, p. 135–150, doi: 10.1016/0012-821X(85)90154-2.

Bralower, T.J., Thomas, D.J., Zachos, J.C., Hirschmann, M.M., Rohl, U., Sigurdsson, H., Thomas, E., and Whitney, D.L., 1997, High-resolution records of late Paleocene thermal maximum and circum-Caribbean volcanism: Is there a causal link?: Geology, v. 25, p. 963–966, doi: 10.1130/0091-7613(1997)025<0963:HRROTL>2.3.CO;2.

Bralower, T.J., Premoli-Silva, I., et al., 2002, Initial reports, Ocean Drilling Program, Leg 198: College Station, Texas, Ocean Drilling Program, 1000 p.

Broecker, W.S., 1997, Will our future ride into the greenhouse be a smooth one?: GSA Today, v. 7, no. 5, p. 1–7.

Broecker, W.S., 1998, Paleocean circulation during the last deglaciation: A bipolar seesaw?: Paleoceanography, v. 13, p. 119–121, doi: 10.1029/97PA03707.

Broecker, W.S., Gerard, R., Ewing, M., and Heezen, B.C., 1960, Natural radiocarbon in the Atlantic Ocean: Journal of Geophysical Research, v. 65, p. 2903–2931.

Burton, K.W., Ling, H.-F., and O'Nions, R.K., 1997, Closure of the Central American Isthmus and its effect on deep-water formation in the North Atlantic: Nature, v. 386, p. 382–385, doi: 10.1038/386382a0.

Christensen, J.N., Halliday, A.N., Godfrey, L.V., Hein, J.R., and Rea, D.K., 1997, Climate and ocean dynamics and the lead isotopic records in Pacific ferromanganese crusts: Science, v. 277, p. 913–918, doi: 10.1126/science.277.5328.913.

DePaolo, D.J., and Wasserburg, G.J., 1976, Nd isotopic variations and petrogenetic models: Geophysical Research Letters, v. 3, p. 248–252.

Elderfield, H., and Greaves, M.J., 1982, The rare earth elements in seawater: Nature, v. 296, p. 214–219, doi: 10.1038/296214a0.

Elderfield, H.R., Upstill-Goddard, J., and Sholkolvitz, E.R., 1990, The rare earth elements in rivers, estuaries, and coastal seas and their significance to the composition of ocean waters: Geochimica et Cosmochimica Acta, v. 54, p. 971–991.

Goldstein, S.L., and Jacobsen, S.B., 1988, Nd and Sr isotope systematics of river water suspended material: Implications for crustal evolution: Earth and Planetary Science Letters, v. 87, p. 249–265, doi: 10.1016/0012-821X(88)90013-1.

Halliday, A.N., Davidson, J.P., Holden, P., Owen, R.M., and Olivarez, A.M., 1992, Metalliferous sediments and the scavenging residence time of

Nd near hydrothermal vents: Geophysical Research Letters, v. 19, p. 761–764.

Jeandel, C., 1993, Concentration and isotopic composition of Nd in the South Atlantic Ocean: Earth and Planetary Science Letters, v. 117, p. 581–591, doi: 10.1016/0012-821X(93)90104-H.

Jones, C.E., Halliday, A.N., Rea, D.K., and Owen, R.M., 1994, Neodymium isotopic variations in the North Pacific modern silicate sediment and the insignificance of detrital REE contributions to seawater: Earth and Planetary Science Letters, v. 127, p. 55–66, doi: 10.1016/0012-821X(94)90197-X.

Jones, C.E., Halliday, A.N., Rea, D.K., and Owen, R.M., 2000, Eolian inputs of lead to the North Pacific: Geochimica et Cosmochimica Acta, v. 64, p. 1405–1416, doi: 10.1016/S0016-7037(99)00439-1.

Levitus, S., 1982, Climatological atlas of the world ocean: National Oceanographic and Atmospheric Administration Professional Paper, v. 13, 173 p.

Ling, H.-F., Burton, K.W., O'Nions, R.K., Kamber, B.S., von Blanckenburg, F., Gibb, A.J., and Hein, J.R., 1997, Evolution of Nd and Pb isotopes in Central Pacific seawater from ferromanganese crusts: Earth and Planetary Science Letters, v. 146, p. 1–12, doi: 10.1016/S0012-821X(96)00224-5.

Martin, E.E., and Haley, B.A., 2000, Fossil fish teeth as proxies for seawater Sr and Nd: Geochimica et Cosmochimica Acta, v. 64, p. 835–847, doi: 10.1016/S0016-7037(99)00376-2.

Martin, E.E., and Scher, H.D., 2004, Preservation of seawater Sr and Nd isotopes in fossil fish teeth: Bad news and good news: Earth and Planetary Science Letters, v. 220, p. 25–39, doi: 10.1016/S0012-821X(04)00030-5.

O'Nions, R.K., Frank, M., von Blanckenburg, F., and Ling, H.-F., 1998, Secular variation of Nd and Pb isotopes in ferromanganese crusts from the Atlantic, Indian, and Pacific Oceans: Earth and Planetary Science Letters, v. 155, p. 15–28, doi: 10.1016/S0012-821X(97)00207-0.

Pettke, T., Halliday, A.N., and Rea, D.K., 2002, Cenozoic evolution of Asian climate and sources of Pacific seawater Pb and Nd derived from eolian dust of sediment core LL44–GPC3: Paleoceanography, v. 17, p. 3-1–3-13.

Piepgras, D.J., and Jacobsen, S.B., 1988, The isotopic composition of neodymium in the North Pacific: Geochimica et Cosmochimica Acta, v. 52, p. 1373–1381, doi: 10.1016/0016-7037(88)90208-6.

Piepgras, D.J., and Wasserburg, G.J., 1982, Isotopic composition of neodymium in waters from the Drake Passage: Science, v. 217, p. 207–214.

Piepgras, D.J., and Wasserburg, G.J., 1987, Rare earth element transport in the western North Atlantic inferred from Nd isotopic observations: Geochimica et Cosmochimica Acta, v. 51, p. 1257–1271, doi: 10.1016/0016-7037(87)90217-1.

Saunders, A.D., Fitton, J.G., Kerr, A.C., Norry, M.J., and Kent, R.W., 1997, The North Atlantic Igneous Province, *in* Mahoney, J.J. and Coffin, M., eds., Large Igneous Provinces: Continental, Oceanic, and Planetary Flood Volcanism: American Geophysical Union Monograph, v. 100, p. 45–93.

Scher, H. and Martin, E.E., 2001, Eocene to Miocene Southern Ocean deep water circulation revealed from fossil fish teeth Nd isotopes: Eos (Transactions, American Geophysical Union), Fall Meeting Supplement, v. 82, Abstract F 639.

Schmitz, W.J., Jr., 1995, On the interbasin-scale thermohaline circulation: Reviews of Geophysics, v. 33, p. 151–173, doi: 10.1029/95RG00879.

Sigurdsson, H., Leckie, R.M., et al., 1997, Initial reports, Ocean Drilling Program, Leg 165: College Station, Texas, Ocean Drilling Program, 1000 p.

Shaw, H.F., and Wasserburg, G.J., 1985, Sm-Nd in marine carbonates and phosphates: Geochimica et Cosmochimica Acta, v. 49, p. 503–518, doi: 10.1016/0016-7037(85)90042-0.

Sholkovitz, E.R., 1993, The geochemistry of rare earth elements in the Amazon River estuary: Geochimica et Cosmochimica Acta, v. 57, p. 2181–2190, doi: 10.1016/0016-7037(93)90559-F.

Staudigel, H., Doyle, P., and Zindler, A., 1985, Sr and Nd isotope systematics in fish teeth: Earth and Planetary Science Letters, v. 76, p. 45–56, doi: 10.1016/0012-821X(85)90147-5.

Tachikawa, K., Jeandel, C., and Roy-Barman, M., 1999, A new approach to the Nd residence time in the ocean: The role of atmospheric inputs: Earth and Planetary Science Letters, v. 170, p. 433–446, doi: 10.1016/S0012-821X(99)00127-2.

Tanaka, T., Togashi, S., Kamioka, H., Amakawa, H., Kagami, H., Hamamoto, T., Yuhara, M., Orihashi, Y., Yoneda, S., Shimizu, H., Kunimaru, T., Takahashi, K., Yanagi, T., Nakano, T., Fujimaki, H., Shinjo, R., Asahara, Y., Tanimizu, M., and Dragusanu, C., 2000, JNdi-1; a neodymium isotopic reference in consistency with LaJolla neodymium: Chemical Geology, v. 168, p. 279–281, doi: 10.1016/S0009-2541(00)00198-4.

Thomas, D.J., 2004, Evidence for deep-water production in the North Pacific Ocean during the early Cenozoic warm interval: Nature, v. 430, p. 65–68.

Thomas, D.J., Bralower, T.J., and Jones, C.E., 2003, Neodymium isotopic reconstruction of late Paleocene–early Eocene thermohaline circulation: Earth and Planetary Science Letters, v. 209, p. 309–322, doi: 10.1016/S0012-821X(03)00096-7.

Tomczak, M., and Godfrey, J.S., 1994, Regional Oceanography: An Introduction: London, Pergamon, 422 p.

Toyoda, K., and Tokonami, M., 1990, Diffusion of rare-earth elements in fish teeth from deep-sea sediments: Nature, v. 345, p. 607–609, doi: 10.1038/345607a0.

Wright, J., Seymour, R.S., and Shaw, H., 1984, REE and Nd isotopes in conodont apatite: Variations with geological age and depositional environment: *in* Clark, D.L., ed., Conondont Biofacies and Provincialism: Geological Society of America Special Paper 196, p. 325–340.

Zachos, J.C., Stott, L.D., and Lohmann, K.C., 1994, Evolution of early Cenozoic marine temperatures: Paleoceanography, v. 9, p. 353–387, doi: 10.1029/93PA03266.

Zachos, J.C., Pagani, M., Sloan, L.C., Thomas, E., and Billups, K., 2001, Trends, rhythms, and aberrations in global climate 65 Ma to present: Science, v. 292, p. 686–693, doi: 10.1126/science.1059412.

Manuscript Accepted by the Society 19 April 2005

Printed in the USA

Geological Society of America
Special Paper 395
2005

Molluscan radiocarbon as a proxy for El Niño–related upwelling variation in Peru

C. Fred T. Andrus
Department of Geological Sciences, University of Alabama, Tuscaloosa, Alabama 35487, USA

Gregory W.L. Hodgins
National Science Foundation–Arizona Accelerator Mass Spectrometry Facility, Department of Physics, University of Arizona, Tucson, Arizona 85721, USA

Daniel H. Sandweiss
Department of Anthropology and Institute for Quaternary and Climate Studies, University of Maine, 120 Alumni Hall, Orono, Maine 04469, USA

Douglas E. Crowe
Department of Geology, University of Georgia, Athens, Georgia 30602, USA

ABSTRACT

Sequential measurements of molluscan radiocarbon are demonstrated to be an effective proxy of seasonal and El Niño–related upwelling variation in coastal Peru. A *Trachycardium procerum* valve from southern Peru was measured through ontogeny for radiocarbon via accelerator mass spectrometry (AMS) as well as $\delta^{13}C$ and $\delta^{18}O$. A specimen collected in 1984 near Casma, Peru (~9.30°S) grew before and during the 1982–1983 El Niño/southern oscillation warm event. Shell morphology recorded El Niño warming as a shallow growth break with subsequent realignment of aragonite crystal microstructure. The presence of this growth pattern indicates that shell material was precipitated during the warm event and that each AMS sample could be independently identified to represent a defined period in the El Niño/southern oscillation cycle. Samples taken from portions of the shell precipitated prior to the El Niño warm event (before the diagnostic growth break) had a mean value of 99.8 percent modern carbon (pMC), with a maximum seasonal range of 2.1 pMC. During warming, as indicated by a negative excursion in $\delta^{18}O$ and the growth break, there was an abrupt increase to 107.9 pMC. Aragonite precipitated near the margin of the valve, after the El Niño/southern oscillation event concluded, had radiocarbon values approaching those present before the growth break. We attribute this radiocarbon distribution to variations in vertical mixing of surface and deeper upwelled water of greater ventilation age. As an El Niño event begins and the thermocline deepens, less deep water reaches the surface. Thus, radiocarbon values in shell precipitated during El Niño appear younger (more positive) relative to non–El Niño periods, which represent periods of more intense upwelling. The results from the modern specimen validate the use of molluscan radiocarbon as a proxy of upwelling conditions related to El Niño/southern oscillation and suggest the utility of similar analysis of more ancient valves in both oceanography and paleoclimatology.

fandrus@geo.ua.edu

Andrus, C.F.T., Hodgins, G.W.L., Sandweiss, D.H., and Crowe, D.E., 2005, Molluscan radiocarbon as a proxy for El Niño–related upwelling variation in Peru, *in* Mora, G., and Surge, D., eds., Isotopic and elemental tracers of Cenozoic climate change: Geological Society of America Special Paper 395, p. 13–20, doi: 10.1130/2005.2395(02).

Keywords: mollusk, radiocarbon, stable oxygen isotope, El Niño, El Niño/southern oscillation, upwelling, thermocline, Peru.

INTRODUCTION

One of the diagnostic physical characteristics of El Niño on the northern coast of Peru is variation in the upwelling of deep, nutrient-rich water. Conditions in this region include strong upwelling and a shallow thermocline in non–El Niño years. When El Niño forms and the thermocline deepens, the source water for upwelling changes from the colder, nutrient-rich sub-thermocline water to nutrient-depleted, warm water originating above the thermocline (see Huyer et al., 1987 and Toggweiler et al., 1991 for a more detailed description). Additionally, the intensity of upwelling may also vary as wind direction and speed change, particularly north of 10°S latitude (Huyer et al., 1987). The net result is that during El Niño, less deep water reaches the surface.

Although upwelling is a central defining characteristic of the El Niño/southern oscillation phenomenon, the instrumental records of it are brief and more ancient proxy data are sparse. Most paleo-upwelling information for the west coast of South America is derived from interpretations of sea surface temperature (SST) proxies (e.g., Andrus et al., 2002) or microfossil-based paleoproductivity estimates from nearby regions (DeVries and Schrader, 1981; Wefer et al., 1983; Loubere, 1999; Marchant et al., 1999; Hebbeln et al., 2002). The potential value of paleo-upwelling data is great. For example, there is a debate concerning the frequency of El Niño and the related mean and seasonal variation in SST in the early to middle Holocene (e.g., Shulmeister and Lees, 1995; Sandweiss et al., 1996, 1997, 2001; DeVries et al., 1997; Wells and Noller, 1997; Thompson et al., 1995; Gagan et al., 1998; Rodbell et al., 1999; Tudhope et al., 2001; Riedinger et al., 2002; Andrus et al., 2002, 2003; Koutavas et al., 2002; Moy et al., 2002; Béarez et al., 2003). Upwelling data could help distinguish the effects of the role of deep and surface water currents in these changes and better define the mean and seasonal upwelling conditions within which El Niño may have operated in the past.

Past upwelling is assessed in some tropical regions by analyzing the radiocarbon content, through ontogeny, of relatively recent corals (e.g., Druffel, 1981, 1982, 1987; Druffel and Griffin, 1993, 1997; Moore et al., 1997; Guilderson et al., 1998, 2000, 2002). The rationale behind this approach is that deeper water contains dissolved inorganic carbon (DIC) that is depleted in ^{14}C relative to surface water due to the length of time the water has been isolated from atmospheric carbon (ventilation age). In an area of temporally variable vertical mixing, such as that caused by El Niño/southern oscillation, organisms that continuously precipitate carbonate skeletons will record any changes in ambient water ^{14}C in their skeletal ^{14}C. If the age of the skeleton can be independently constrained, for example through analysis of incremental growth and/or through independent age dating, and the ^{14}C age of the carbonate is calculated, the difference in ages will be a function of the degree of vertical mixing and ventilation age. Sequential analyses of such a carbonate would track how the water column changes through time. For example, during an El Niño event, as the deep water with a relatively old ventilation age is displaced by surface water, this variation would be recorded in carbonates growing throughout the change.

Nuclear bomb–derived radiocarbon currently enhances the difference between the apparent ages of deep and surface water. The enrichment in ^{14}C in surface waters has been used as a large-scale oceanic current tracer by marking the temporal and spatial progression of this anthropogenic radiocarbon in water and coral (see Östlund and Stuiver, 1980; Broecker et al., 1985; McNichol et al., 2000; Key et al., 1996, 2002). In the absence of bomb-derived radiocarbon, as would be the case in paleo-upwelling proxies, the radiocarbon difference between surface- and deep-ocean currents would provide a measure of only natural ventilation age variation. This method could be applied to ancient carbonates in Peru to elucidate the history of El Niño/southern oscillation–related upwelling patterns.

As the coast of Peru is presently dominated by temperate water, no coral exist that are suitable for the analyses described above, however mollusks represent a viable alternative. Mollusk shells are a frequently used and reliable source of ^{14}C data for purposes of constraining ΔR, the local variation in reservoir effect (e.g., Taylor and Berger, 1967; Southon et al., 1990, 1992), shell growth rate (Turekian et al., 1982; Landman et al., 1988), and measurement of recent variation in the eastern equatorial Pacific thermocline (Toggweiler et al., 1991), in addition to multiple applications of age dating archaeological materials (e.g., Sandweiss et al., 1989; Kennett et al., 2002). The abundance of a wide variety of mollusk species in archaeological sites on the Pacific coast of South America (e.g., Sandweiss et al., 1996, 2001) and the presence of fossil assemblages of similar or more ancient age (e.g., Hsu et al., 1989) create the possibility of measuring changes in deep-water upwelling from the late Pleistocene through the Holocene. By measuring changes in ΔR through time as evidenced by comparisons between shell radiocarbon and known age dates, a broad description of variation in upwelling could be created, and by analyzing individual shells at high resolution through ontogeny, seasonal variation in upwelling at different time periods could be measured.

The research we describe here is a modern analogue validation of the application of molluscan radiocarbon as a proxy for El Niño/southern oscillation–related upwelling. The effects of the strong 1982–1983 El Niño on shell radiocarbon and $\delta^{18}O$ are described.

SPECIES DESCRIPTION

The species selected for analysis is the cockle *Trachycardium procerum* (Mollusca: Bivalvia: Cardiidae), sometimes locally called mule's hoof (*pata de mula*) (Fig. 1). It ranges in the eastern Pacific from Mexico to Chile (Olsson, 1961). This

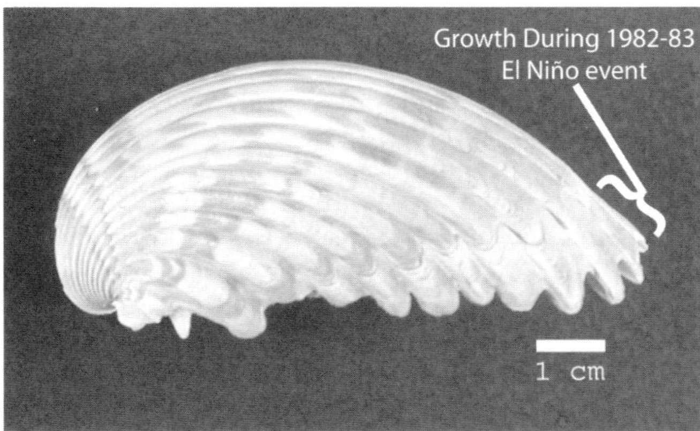

Figure 1. Profile view of *T. procerum* valve. The growth break is indicated by the left side of the bracket. Subsequent growth underneath the bracket displays a change in valve curvature caused by realignment of aragonite microstructure.

species is present in well-dated archaeological contexts, modern collections, and in deposits from the last several interglacial epochs, so it could be a useful climate proxy for the Holocene and later Pleistocene. It is common in northern Peru in littoral and embayed environments where it is a shallow infaunal filter feeder favoring sandy substrates. *T. procerum* produces relatively large (>8–10 cm) ribbed valves composed of aragonite.

The growth patterns in the valves in this analysis were described in detail by Rollins et al. (1986, 1987). The samples were collected in 1984 and lived throughout the strong 1982–1983 El Niño event. Rollins et al. (1987) noted that the incremental microstructure changed in the aragonite precipitated during El Niño. A broad and shallow growth break occurred during El Niño warming. Analysis of subdaily increments by Rollins et al. (1987) indicated that the growth break occurred in April and June in the two shells described. The orientation of the laminae changed in subsequent growth in these individuals. Microincrements in *T. procerum* ordinarily have an angle of inclination to the valve surface of ~35°, but during El Niño the increments are precipitated at ~10° to the valve surface. This microstructural change resulted in a readily apparent "flattening" of shell curvature indicative of material precipitated during El Niño (Fig. 1). This anomaly is present in all of the shells examined in the modern collection but was more prominent in some individuals than others.

Population studies along coastal Peru have shown that *T. procerum* survives El Niño events, whereas heat-intolerant species of mollusk, such as *Mesodesma donacium,* suffer near complete mortality north of 12°S (Rollins et al., 1986, 1987; Sandweiss et al., 2001). Moreover, the growth anomaly found in this *T. procerum* indicates that this species continues to grow during periods of temperature stress: a significant quantity of shell carbonate was precipitated during the 1982–1983 El Niño event. Based on the timing of the El Niño–related growth break as documented by Rollins et al. (1987), this organism did not become stressed until several months after the onset of the warm event. Therefore the shell chemistry should reflect related local environmental variation through this time. The presence of the growth "scar" provided an unambiguous independent measure of the timing of shell precipitation relative to the progression of the El Niño event.

METHODS

The *T. procerum* valve analyzed in this research (2TP4–2) was collected by commercial divers near Los Chimus (~9.30°), between Chimbote (9.07°S) and Casma (9.44°S) (Fig. 2). Soft tissue was removed from the valve soon after collection and the valve was stored dry until analysis. Prior to sampling, the outer surface of the valve was cleaned, abraded by a wire brush, and washed in distilled water to limit surface contamination.

Thirty samples were milled from the outer surface along transects parallel with macroscopic growth increments. Sampling followed ontogeny and the location of each sample relative to growth structures was noted in addition to measuring the distance of the sample from the hinge along the longest axis of growth. Each transect was ~0.5 mm wide and <0.5 mm deep. Near the hinge area and near the outer edge (where abrasion had

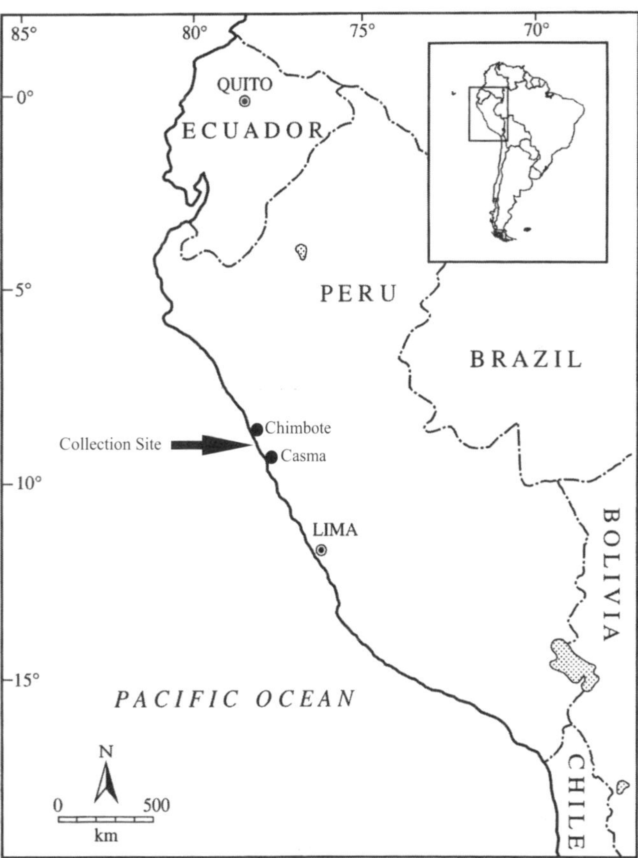

Figure 2. Map showing location of collection site in coastal Peru with locations of nearby cities.

damaged the shell) multiple transects (usually 2–3) were milled to generate the ~3–4 mg of powdered sample needed for AMS measurement. Only one transect was required in other areas of the shell. Smaller samples (<1.0 mg) were taken from the same transects for $\delta^{13}C$ and $\delta^{18}O$ analyses. CO_2 was extracted from the samples following conventional phosphoric acid techniques and was cryogenically purified.

Samples for ^{14}C analysis were reduced to graphite using an iron catalyst following the method of Vogel et al. (1984). Graphite $^{14}C/^{13}C$ ratios were measured by AMS at the University of Georgia AMS Laboratory at the Center for Applied Isotope Studies. The calculations for determination of pMC were based upon ratios measured from oxalic acids I and II standards and ^{14}C-dead process blank samples. The uncorrected pMC of the samples was calculated according to equations presented in Stuiver and Polach (1977) and Donahue et al. (1990).

$\delta^{13}C$ and $\delta^{18}O$ values were measured on a Finnigan MAT 252 isotope ratio mass spectrometer (IRMS) at the University of Georgia Department of Geology Stable Isotope Laboratory. All data were reported in parts per mil (‰) relative to the Vienna Peedee belemnite standard with precision calculated as ~0.07‰ (1σ) for both $\delta^{18}O$ and $\delta^{13}C$ based on analysis of the University of Georgia working standard Iceland Spar (Landis, 1983). These $\delta^{13}C$ values were used for the isotopic fractionation correction in the calculations of $\Delta^{14}C$.

RESULTS

Of the thirty samples that were milled from the valve, each was measured for $\delta^{13}C$ and $\delta^{18}O$, while 29 were measured for $\Delta^{14}C$, with one AMS sample (#4, near the hinge) being lost during sample preparation. The $\Delta^{14}C$ data generated from the *T. procerum* shell are plotted in Figure 3 and reflect $\delta^{13}C$ corrections. The $\delta^{18}O$ data are plotted in Figure 4. The $\delta^{13}C$ data are plotted in Figure 5. Precision (1σ) is displayed by the *y*-axis error bars on each graph. The distance of each sample relative to the hinge is noted on the *x*-axis of each graph. To facilitate comparison between the profiles, all are plotted together in Figure 6.

The $\Delta^{14}C$ and $\delta^{18}O$ values early in ontogeny are significantly different than those in the latter half of growth. The mean $\Delta^{14}C$ value is 101.3 pMC (±2.7; 1σ). The mean $\delta^{18}O$ value is –1.7‰ (±0.4; 1σ). Using the growth anomaly created by the 1983–1984 El Niño event as a benchmark, the mean "pre–El Niño" $\Delta^{14}C$ value is 99.8 (±1.1; 1σ), and the mean "post–El Niño" value is

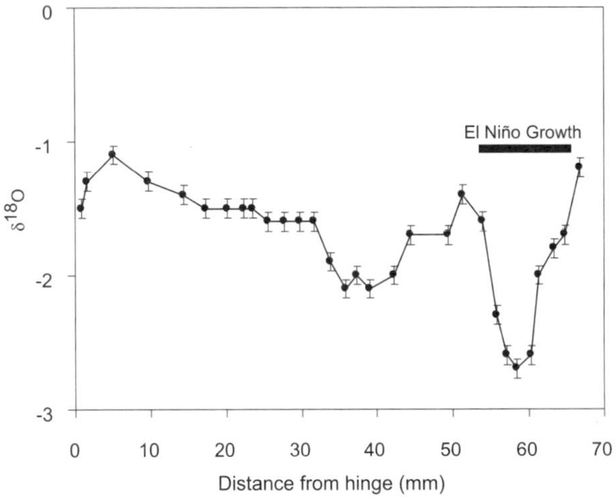

Figure 4. $\delta^{18}O$ profile of *T. procerum* reported in parts per mil (‰) relative to the Vienna Peedee belemnite (VPDB) standard through ontogeny (left to right). *y*-axis error bars represent precision (0.07‰; 1σ). The negative excursion in $\delta^{18}O$ on the right side of the graph occurred in aragonite precipitated during the 1982–1983 El Niño event.

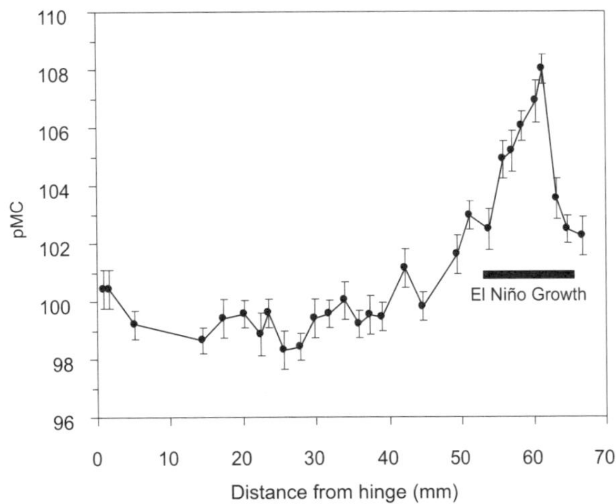

Figure 3. $\Delta^{14}C$ profile of *T. procerum* reported in percent modern carbon (pMC) through ontogeny (left to right); *y*-axis error bars represent precision (1σ). The positive excursion in $\Delta^{14}C$ on the right side of the graph occurred in aragonite precipitated during the 1982–1983 El Niño event.

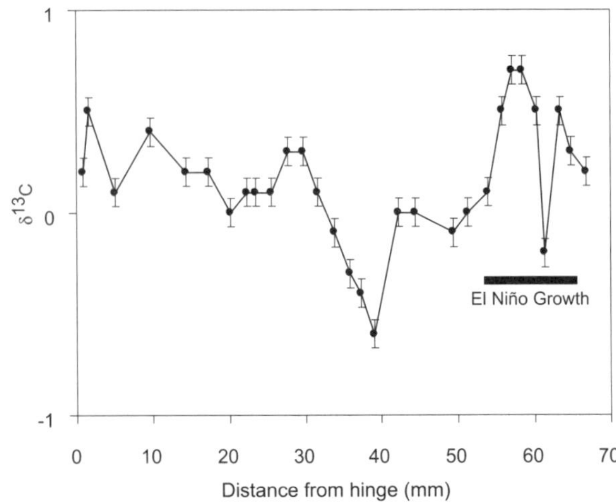

Figure 5. $\delta^{13}C$ profile of *T. procerum* reported in parts per mil (‰) relative to the Vienna Peedee belemnite (VPDB) standard through ontogeny (left to right). *y*-axis error bars represent precision (0.07‰; 1σ). Aragonite precipitated during the 1982–1983 El Niño event is identified on the right side of the graph.

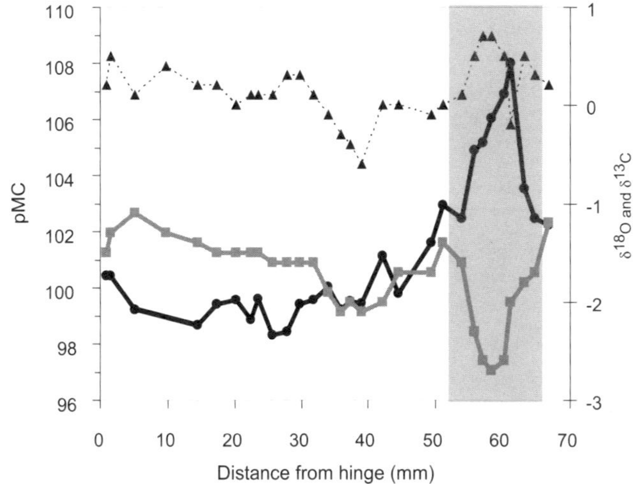

Figure 6. *T. procerum* Δ^{14}C (circles, dark solid line), δ^{18}O (gray squares, gray line), and δ^{13}C (triangles, dashed line) profiles plotted for comparison. Primary (left) *y*-axis is Δ^{14}C in percent modern carbon (pMC). Secondary (right) *y*-axis is δ^{18}O and δ^{13}C in parts per mil (‰) relative to the Vienna Peedee belemnite (VPDB). *x*-axis represents distance from the hinge in millimeters following ontogeny (left to right). Transparent gray bar represents aragonite precipitated during the 1982–1983 El Niño event based on the shell growth anomaly.

104.6 (±2.1; 1σ). The mean "pre–El Niño" δ^{18}O value is –1.6‰ (±0.5; 1σ), and the mean "post–El Niño" value is –2.1‰ (±0.1; 1σ). The mean δ^{13}C value was 0.1‰ (±0.3; 1σ).

Both Δ^{14}C and δ^{18}O profiles followed a roughly sinusoidal pattern; however, the two curves were out of phase. The greatest Δ^{14}C value, 107.9 pMC, was measured in carbonate precipitated during the El Niño, while the most negative δ^{18}O value, –2.7‰, was also from carbonate precipitated during this warm event. The most positive Δ^{14}C value was not exactly synchronous with the most negative δ^{18}O value. The Δ^{14}C profile peaked at 107.9 pMC at 61 mm from the hinge, while the most negative δ^{18}O value, –2.7‰, occurred 58 mm from the hinge. The El Niño growth break occurred prior to this most negative sample at 53.9 mm from the hinge. At the margin of the valve, Δ^{14}C and δ^{18}O values became more similar to those measured earlier in ontogeny.

DISCUSSION

The *T. procerum* Δ^{14}C profile (Figs. 3 and 6) follows the pattern predicted to occur in carbonates formed during the El Niño cycle in Peru. During an El Niño along this coast, the thermocline deepens and the source of upwelling water changes from the colder, nutrient-rich (containing older, ^{14}C-depleted DIC) subthermocline water to nutrient-poor, warm (containing younger, ^{14}C-enriched DIC) water largely originating above the thermocline. Upwelling intensity may diminish as well. Less of the deep, ^{14}C-depleted water reaches the surface during El Niño, thus carbonates precipitated during this event are ^{14}C-enriched relative to times before and after the event.

A significant positive excursion in Δ^{14}C occurs coincident with carbonate precipitated during El Niño, indicating the encroachment of ^{14}C-enriched surface water. Prior to this excursion, the mean Δ^{14}C value is 99.8 pMC with a maximum seasonal range of 2.1 pMC. Huyer et al. (1987) suggested that before, during, and after the 1982–1983 El Niño the source depth of upwelled water was between 50 and 100 m. In order to test this hypothesis with the *T. procerum* data, detailed Δ^{14}C data from the entire water column closely adjacent to the collection site are necessary. Unfortunately such data are not available. As part of the World Ocean Circulation Experiment (Key et al., 2002; http://www.woce.org) in 1993, water samples were measured nearly 800 km offshore (longitude ~86°W) near the latitude of the *T. procerum* collection site. Unfortunately, this distance between the shell collection site and the World Ocean Circulation Experiment data is too great to ensure a precise and meaningful determination of the depth of upwelling, especially considering the trend for the thermocline to be shallower near the coast.

Another complicating factor in ^{14}C comparisons between the shell and the World Ocean Circulation Experiment data is that the shell was collected almost a decade before the sampling, thus bomb radiocarbon may not have penetrated as deeply into the water column during the life of the *T. procerum.* Unfortunately, the earlier Geochemical Ocean Sections Study did not sample proximal to the coast of Peru (Östlund and Stuiver, 1980), so no estimate of local change in depth of bomb-radiocarbon penetration can be confidently made.

Although a precise comparison between the shell and water DIC radiocarbon is not possible at this point due to limited local seawater data, the mean pre–El Niño Δ^{14}C value in the *T. procerum* is consistent with the shell growing in a mixture of deeper water and bomb-enriched surface water. As the surface water component in the mixture begins to dominate during El Niño, the shell becomes ^{14}C-enriched to a maximum value of 107.9 pMC (Figs. 3 and 6).

Clearly, bomb-generated radiocarbon in the surface ocean exaggerates the amplitude of seasonal and El Niño oscillations in Δ^{14}C. In a prebomb shell, as would be encountered in archaeological or fossil deposits, the difference in age between El Niño and non–El Niño growth would be a function of upwelling and ventilation age of the deep water component. In order to determine the pattern of El Niño/southern oscillation and seasonal variability without the influence of anthropogenic ^{14}C, a prebomb valve captured at a known time must be sampled. However, Toggweiler et al. (1991) estimated that the Δ^{14}C in prebomb upwelled water off Peru would range between 91.9 pMC and 92.8 pMC, thus we feel that differences between surface and upwelled currents would be readily detectable even in the absence of bomb radiocarbon.

The overall pattern in the δ^{18}O profile (negative values in carbonate precipitated during El Niño) is consistent with measured contemporaneous local SST variation, but the maximum amplitude of variation caused by El Niño seems attenuated relative to measured SST range (Figs. 4 and 6). In the molluscan aragonite δ^{18}O/temperature equation (Grossman and Ku, 1986), a change of 0.21‰ is equivalent to 1 °C (assuming stable $\delta^{18}O_{water}$). The mean SST at Chimbote between 1980 and the onset of the 1982–1983 El Niño is ~20 °C and it rose by ~8 °C during the warmest phase of the event (based on mean monthly data: IRI/LDEO Climate Data Library, online at http://ingrid.ldgo.columbia.edu/). This variation would translate to a ~1.7‰ shift in molluscan aragonite $\delta^{18}O$ from mean values to those during El Niño. The variation between the mean $\delta^{18}O$ prior to El Niño growth measured in the shell and the most negative value measured during El Niño growth is 1.1‰. Several observations account for this 0.6‰ difference between Chimbote SST and the shell $\delta^{18}O$ data.

The warm SSTs of the 1982–1983 event persisted for ~12 months, and the associated negative anomaly in $\delta^{18}O$ was measured in 8 samples. Thus the mean time average of the shell samples is ~1.5 months during this period of growth. Time averaging therefore contributes to some of the difference between the instrumental SST record and temperatures predicted from $\delta^{18}O$ measured in the *T. procerum* shell. Additionally, the distance between the collection site and the data station may account for some differences in the SST records. The shallow endobenthic habitat of the species would also be a factor in insulating the valve from exposure to the greatest temperature extremes. There is a growth break of an unknown duration near the time of peak El Niño SST warming (Rollins et al., 1986) that may diminish the maximum amplitude of $\delta^{18}O$ recorded in the shell during El Niño. More analyses of modern specimens of *T. procerum* collected from monitored locations in conjunction with detailed SST data are necessary before assessments of absolute temperatures can be made with confidence. Nevertheless, the data seem robust in terms of relative variation in temperature throughout the period of growth of an individual valve. The negative excursion in $\delta^{18}O$ in the profile is diagnostic of El Niño and is therefore valuable as a potential proxy as suggested by Rollins et al. (1987), Perrier et al. (1994), and Andrus et al. (2003).

Compared to $\Delta^{14}C$ and $\delta^{18}O$, relatively little variation in $\delta^{13}C$ occurred throughout the life of the shell (Figs. 5 and 6). The maximum range in $\delta^{13}C$ is 1.3‰. The mean values in the aragonite precipitated during the El Niño event are more positive than pre–El Niño shell material by 0.3‰. This general pattern is consistent with differences in ambient $\delta^{13}C$ from upwelling currents versus surface waters. Typically in upwelled waters, the DIC available to growing mollusks is relatively depleted in ^{13}C compared with surface waters, due to increased productivity (Killingley and Berger, 1979). As surface currents displace deeper upwelling during El Niño, the local waters off Peru contain more ^{13}C-enriched DIC due to decreased productivity, and this is reflected in the aragonite precipitated in the *T. procerum.* However, given that the maximum deviation in $\delta^{13}C$ values associated with El Niño (0.7‰) is less than deviations earlier in ontogeny (0.9‰), the $\delta^{13}C$ profile does not provide a robust measure of El Niño.

The combined $\Delta^{14}C$ and $\delta^{18}O$ profiles (Fig. 6) are an unambiguous diagnostic indicator of El Niño events. It is difficult to construct a reasonable alternative causal hypothesis for the large and simultaneous positive $\Delta^{14}C$ excursion and negative $\delta^{18}O$ excursion in the *T. procerum,* and therefore this dual proxy can be employed to detect El Niño in ancient shells. Additionally, the presence of the shell growth anomaly could also contribute to defining past El Niño events. Other stressors could create similar growth anomalies, and events weaker than the large 1982–1983 event may not produce any obvious growth variation. Nevertheless, the presence of such potential El Niño/southern oscillation markers in shells from archaeological middens or fossil deposits would suggest that detailed $\Delta^{14}C$ and $\delta^{18}O$ analyses would define an El Niño event. This may be a profitable strategy for analysis of the upwelling and SST structure of past events.

The $\Delta^{14}C$ data presented here suggest caution in the interpretation of estimated ΔR or age dates derived from whole-shell analyses in regions subject to El Niño/southern oscillation–related upwelling. Had the *T. procerum* shell analyzed through ontogeny in this study been measured for age data via whole-shell analyses, a significant error would have been introduced through short-term variation in ΔR. Similarly, shells collected shortly after El Niño events that are used to determine ΔR may provide misleading data due to upwelling variation that occurred at some point earlier in ontogeny. In order to avoid these potential difficulties, mollusks analyzed for such purposes should be sampled sequentially to control for large short-term variations in $\Delta^{14}C$ caused by upwelling/thermocline shifts. Sampling resolution need not be as fine as in this study, but should be conducted at a temporal resolution such that interannual scale change is detectable. Twelve-month temporal resolution in sampling may be sufficient to detect such events.

CONCLUSIONS

Molluscan radiocarbon profiles may serve as proxies of past upwelling related to El Niño/southern oscillation in coastal Peru. Fine temporal resolution analyses, as performed in this study, contain data pertinent to defining seasonal and mean upwelling conditions and ventilation ages. Because *T. procerum* and some other mollusk species survive the temperature extremes associated with the El Niño/southern oscillation cycle, the coupled $\Delta^{14}C$ and $\delta^{18}O$ profiles from the valves of these organisms would contain evidence of particular past El Niño events. If longer-lived species can be identified and found in contexts, archaeological or otherwise, that provide independent age dates it may be possible to reconstruct El Niño frequency histories. In the absence of long-lived species for analysis, data from shorter time windows will nevertheless be a source of valuable insight concerning general upwelling conditions at the time of shell growth. Additionally, comparisons between shell $\Delta^{14}C$ values and associated terrestrial age dates will permit reconstruction of past variation in ventilation age, as advocated and conducted by Stuiver and Braziunas (1993) and Southon et al. (1990, 1992, 1995). As upwelling is a central defining factor in El Niño/southern oscillation, all of these potential applications may significantly contribute to better understanding of this global climate phenomenon.

ACKNOWLEDGMENTS

We thank Mark Roberts, Doug Dvoracek of the University of Georgia Center for Applied Isotope Studies, and Bill McClain from the Department of Geology Stable Isotope Laboratory. This research was funded in part by the National Science Foundation grant ATM-0082213 (D.E.C.) and Department of Energy grant DE-FC09-96SR18546 (C.F.T.A.).

REFERENCES CITED

Andrus, C.F.T., Crowe, D.E., Sandweiss, D.H., Reitz, E.J., and Romanek, C.S., 2002, Otolith $\delta^{18}O$ record of mid-Holocene sea surface temperatures in Peru: Science, v. 295, p. 1508–1511, doi: 10.1126/science.1062004.

Andrus, C.F.T., Crowe, D.E., Sandweiss, D.H., Reitz, E.J., Romanek, C.S., and Maasch, K.A., 2003, Response to comment on "Otolith $\delta^{18}O$ record of mid-Holocene sea surface temperatures in Peru": Science, v. 299, p. 203b, doi: 10.1126/science.1077525.

Béarez, P., DeVries, T.J., and Ortlieb, L., 2003, Comment on "Otolith $\delta^{18}O$ record of mid-Holocene sea surface temperatures in Peru": Science, v. 299, p. 203a, doi: 10.1126/science.1076173.

Broecker, W.S., Peng, T.-H., Östlund, H.G., and Stuiver, M., 1985, The distribution of bomb radiocarbon in the ocean: Journal of Geophysical Research–Oceans, v. 90, p. 6953–6970.

DeVries, T.J., and Schrader, H., 1981, Variation of upwelling/oceanic conditions during the latest Pleistocene through Holocene off the central Peruvian coast: A diatom record: Marine Micropaleontology, v. 6, p. 157–167, doi: 10.1016/0377-8398(81)90003-7.

DeVries, T.J., Ortlieb, L., Diaz, A., Wells, L.E., and Hillaire-Marcel, C., 1997, Determining the early history of El Niño—Comment: Science, v. 276, p. 965–966, doi: 10.1126/science.276.5314.965.

Donahue, D.J., Linick, T.W., and Jull, A.J.T., 1990, Isotope-ratio and background corrections for accelerator mass spectrometry radiocarbon measurements: Radiocarbon, v. 32, p. 135–142.

Druffel, E.R.M., 1981, Radiocarbon in annual coral rings from the eastern tropical Pacific Ocean: Geophysical Research Letters, v. 8, p. 59–62.

Druffel, E.R.M., 1982, Banded corals: Changes in oceanic carbon-14 during the Little Ice Age: Science, v. 218, p. 13–19.

Druffel, E.R.M., 1987, Bomb radiocarbon in the Pacific: Annual and seasonal timescale variations: Journal of Marine Research, v. 45, p. 667–698.

Druffel, E.R.M., 1997, Geochemistry of corals: Proxies of past ocean chemistry, ocean circulation, and climate: Proceedings of the National Academy of Sciences of the United States of America, v. 94, p. 8354–8361, doi: 10.1073/pnas.94.16.8354.

Druffel, E.R.M., and Griffin, S., 1993, Large variation of surface ocean radiocarbon: Evidence of circulation changes in the southwestern Pacific: Journal of Geophysical Research, v. 98, p. 20,249–20,259.

Gagan, M.K., Ayliffe, L.K., Hopley, D., Cali, J.A., Mortimer, G.E., Chappell, J., McCulloch, M.T., and Head, M.J., 1998, Temperature and surface-ocean water balance of the mid-Holocene tropical Western Pacific: Science, v. 279, p. 1014–1018, doi: 10.1126/science.279.5353.1014.

Grossman, E.L., and Ku, T.-L., 1986, Oxygen and carbon isotope fractionation in biogenic aragonite: Temperature effects: Chemical Geology, v. 59, p. 59–74, doi: 10.1016/0009-2541(86)90044-6.

Guilderson, T.P., Schrag, D.P., Kashgarian, M., and Southon, J.R., 1998, Radiocarbon variability in the western equatorial Pacific inferred from a high-resolution coral record from Nauru Island: Journal of Geophysical Research–Oceans, v. 103, p. 24,641–24,650.

Guilderson, T.P., Schrag, D.P., Goddard, E., Kashgarian, M., Wellington, G.M., and Linsley, B.K., 2000, Southwest subtropical Pacific surface water radiocarbon in a high-resolution coral record: Radiocarbon, v. 42, p. 249–256.

Guilderson, T.P., Kashgarian, M., and Schrag, D.P., 2002, Biogeochemical proxies in scleractinian corals used to reconstruct ocean circulation, Study of Environmental Change using Isotope Techniques: Vienna, Austria, International Atomic Energy Agency, p. 130–139.

Hebbeln, D., Marchant, M., and Wefer, G., 2002, Paleoproductivity in the southern Peru–Chile current through the last 33,000 yr: Marine Geology, v. 186, p. 487–504, doi: 10.1016/S0025-3227(02)00331-6.

Hsu, J.T., Leonard, E.M., and Wehmiller, J.F., 1989, Aminostratigraphy of Peruvian and Chilean Quaternary marine terraces: Quaternary Science Reviews, v. 8, p. 255–262, doi: 10.1016/0277-3791(89)90040-1.

Huyer, A., Smith, R.L., and Paluszkiewicz, T., 1987, Coastal upwelling off Peru during normal and El Niño times, 1981–1984: Journal of Geophysical Research, v. 92, p. 14,297–14,307.

Kennett, D.J., Ingram, L.B., Southon, J.R., and Wise, K., 2002, Differences in ^{14}C age between stratigraphically associated charcoal and marine shell from the Archaic Period site of Kilometer 4, southern Peru: Old wood or old water?: Radiocarbon, v. 44, p. 53–58.

Key, R.M., Quay, P.D., Jones, G., McNichol, A.P., von Reden, K.F., and Schneider, R.J., 1996, WOCE AMS radiocarbon.1. Pacific Ocean results (P6, P16 and P17): Radiocarbon, v. 38, p. 425–518.

Key, R.M., Quay, P.D., Schlosser, P., McNichol, A.P., von Reden, K.F., Schneider, R.J., Elder, K.L., Stuiver, M., and Östlund, H.G., 2002, WOCE radiocarbon IV: Pacific Ocean results; P10, P13N, P14C, P18, P19 & S4P: Radiocarbon, v. 44, p. 239–392.

Killingley, J.S., and Berger, W.H., 1979, Stable isotopes in a mollusk shell: Detection of upwelling events: Science, v. 205, p. 186–188.

Koutavas, A., Lynch-Stieglitz, J., Marchitto, T.M., Jr., and Sachs, J.P., 2002, El Niño–like pattern in Ice Age tropical Pacific sea surface temperature: Science, v. 297, p. 226–230, doi: 10.1126/science.1072376.

Landis, G.P., 1983, Harding Iceland spar: A new δ18O-δ13C carbonate standard for hydrothermal minerals: Isotope Geoscience, v. 1, p. 91–94.

Landman, N.H., Druffel, E.R.M., Cochran, J.K., Donahue, D.J., and Jull, A.J.T., 1988, Bomb-produced radiocarbon in the shell of the chambered nautilus: Rate of growth and age at maturity: Earth and Planetary Science Letters, v. 89, p. 28–34, doi: 10.1016/0012-821X(88)90030-1.

Loubere, P., 1999, A multiproxy reconstruction of biological productivity and oceanography in the eastern equatorial Pacific for the past 30,000 years: Marine Micropaleontology, v. 37, p. 173–198, doi: 10.1016/S0377-8398(99)00013-4.

Marchant, M., Hebbeln, D., and Wefer, G., 1999, High resolution planktic foraminiferal record of the last 13,300 years from the upwelling area off Chile: Marine Geology, v. 161, p. 115–128, doi: 10.1016/S0025-3227(99)00041-9.

McNichol, A.P., Schneider, R.J., von Reden, K.F., Gagnon, A.R., Elder, K.L., Key, R.M., and Quay, P.D., 2000, Ten years after—The WOCE AMS radiocarbon program: Nuclear Instruments & Methods in Physics Research Section B—Beam Interactions with Materials and Atoms, v. 172, p. 479–484, doi: 10.1016/S0168-583X(00)00093-8.

Moore, M.D., Schrag, D.P., and Kashgarian, M., 1997, Coral radiocarbon constraints on the source of the Indonesian throughflow: Journal of Geophysical Research–Oceans, v. 102, p. 12,359–12,365.

Moy, C.M., Seltzer, G.O., Rodbell, D.T., and Anderson, D.M., 2002, Variability of El Niño/Southern Oscillation activity at millennial timescales during the Holocene epoch: Nature, v. 420, p. 162–165, doi: 10.1038/nature01194.

Olsson, A.A., 1961, Mollusks of the Tropical Eastern Pacific particularly from the southern half of the Panamic-Pacific faunal province (Panama to Peru) Panamic-Pacific Pelecypoda: Ithaca, New York, Paleontological Research Institute, 574 p.

Östlund, H.G., and Stuiver, M., 1980, GEOSECS Pacific radiocarbon: Radiocarbon, v. 22, p. 25–53.

Perrier, C., Hillaire-Marcel, C., and Ortlieb, L., 1994, Paléogéographie littorale et enregistrement isotopique (^{13}C, ^{18}O) d'événements de type El Niño par les mollusques Holocènes et récents du nordouest Péruvien: Géographie Physique et Quaternaire, v. 48, p. 23–38.

Riedinger, M.A., Steinitz-Kannan, M., Last, W.M., and Brenner, M., 2002, A ~6100 ^{14}C yr record of El Niño activity from the Galápagos Islands: Journal of Paleolimnology, v. 27, p. 1–7, doi: 10.1023/A:1013514408468.

Rodbell, D.T., Seltzer, G.O., Anderson, D.M., Abbott, M.B., Enfield, D.B., and Newman, J.H., 1999, An ~15,000-year record of El Niño–driven alluviation in southwestern Ecuador: Science, v. 283, p. 516–520, doi: 10.1126/science.283.5401.516.

Rollins, H.B., Sandweiss, D.H., and Rollins, J.C., 1986, Effect of the 1982–1983 El Niño on bivalve mollusks: National Geographic Research, v. 2, p. 106–112.

Rollins, H.B., Sandweiss, D.H., Brand, U., and Rollins, J.C., 1987, Growth increment and stable isotope analysis of marine bivalves: Implications for the geoarchaeological record of El Niño: Geoarchaeology, v. 2, p. 181–197.

Sandweiss, D.H., Richardson, J.B., Reitz, E.J., Hsu, J.T., and Feldman, R.A., 1989, Early maritime adaptations in the Andes: Preliminary studies at the Ring Site, *in* Rice, D., Stanish, C., and Scar, P., eds., Ecology, Settlement, and History in the Osmore Drainage, Peru, Volume 545(1), BAR International Series, p. 34–84.

Sandweiss, D.H., Richardson, J.B., III, Reitz, E.J., Rollins, H.B., and Maasch, K.A., 1996, Geoarchaeological evidence from Peru for a 5000 years B.P. onset of El Niño: Science, v. 273, p. 1531–1533.

Sandweiss, D.H., Richardson, J.B., Reitz, E.J., Rollins, H.B., and Maasch, K.A., 1997, Determining the early history of El Niño—Response: Science, v. 276, p. 966–967.

Sandweiss, D.H., Maasch, K.A., Burger, R.L., Richardson, J.B., Rollins, H.B., and Clement, A.C., 2001, Variation in Holocene El Niño frequencies: Climate records and cultural consequences in ancient Peru: Geology, v. 29, p. 603–606, doi: 10.1130/0091-7613(2001)029<0603:VIHENO>2.0.CO;2.

Shulmeister, J., and Lees, B.G., 1995, Pollen evidence from tropical Australia for the onset of an ENSO–dominated climate at C 4000 BP: The Holocene, v. 5, p. 10–18.

Southon, J.R., Nelson, D.E., and Vogel, J.S., 1990, A record of past ocean-atmosphere radiocarbon differences from the northeast Pacific: Paleoceanography, v. 5, p. 197–206.

Southon, J.R., Nelson, D.E., and Vogel, J.S., 1992, The determination of past ocean-atmosphere radiocarbon differences, *in* Bard, E., and Broecker, W.S., eds., The Last Deglaciation: Absolute and Radiocarbon Chronologies, Volume 12: NATO ASI Series: Berlin, Springer-Verlag, p. 219–227.

Southon, J.R., Rodman, A.O., and True, D., 1995, A comparison of marine and terrestrial radiocarbon ages from northern Chile: Radiocarbon, v. 37, p. 389.

Stuiver, M., and Braziunas, T.F., 1993, Modeling atmospheric ^{14}C influences and ^{14}C ages of marine samples to 10,000 BC: Radiocarbon, v. 35, p. 137–189.

Stuiver, M., and Polach, H.A., 1977, Discussion: Reporting of ^{14}C data: Radiocarbon, v. 19, p. 355–363.

Taylor, R.E., and Berger, R., 1967, Radiocarbon content of marine shells from the Pacific coasts of Central and South America: Science, v. 158, p. 1180–1182.

Thompson, L.G., Mosley-Thompson, E., Davis, M.E., Lin, P.N., Henderson, K.A., Cole-Dai, J., Bolzan, J.F., and Liu, K.B., 1995, Late glacial stage and Holocene tropical ice core records from Huascarán, Peru: Science, v. 269, p. 46–50.

Toggweiler, J.R., Dixon, K., and Broecker, W.S., 1991, The Peru upwelling and the ventilation of the South Pacific thermocline: Journal of Geophysical Research, v. 96, p. 20,467–20,497.

Tudhope, A.W., Chilcott, C.P., McCulloch, M.T., Cook, E.R., Chappell, J., Ellam, R.M., Lea, D.W., Lough, J.M., and Shimmield, G.B., 2001, Variability in the El Nino–Southern Oscillation through a glacial-interglacial cycle: Science, v. 291, p. 1511–1517, doi: 10.1126/science.1057969.

Turekian, K.K., Cochran, J.K., Nozaki, Y., Thompson, I., and Jones, D.S., 1982, Determination of shell deposition rates of Arctica islandica from the New York Bight using natural ^{228}Ra and ^{228}Th and bomb-produced ^{14}C: Limnology and Oceanography, v. 27, p. 737–741.

Vogel, J.S., Southon, J.R., Nelson, D.E., and Brown, T.A., 1984, Performance of catalytically condensed carbon for use in accelerator mass spectrometry: Nuclear Instruments & Methods in Physics Research. Section B, Beam Interactions with Materials and Atoms, v. 5, p. 289–293, doi: 10.1016/0168-583X(84)90529-9.

Wefer, G., Dunbar, R.B., and Suess, E., 1983, Stable isotopes of foraminifers off Peru recording high fertility and changes in upwelling history, *in* Thiede, J., and Suess, E., eds., Coastal Upwelling: Its Sediment Record. Part B: Sedimentary Records of Ancient Coastal Upwelling: New York, Plenum Press, p. 295–308.

Wells, L.E., and Noller, J., S., 1997, Determining the early history of El Niño—Comment: Science, v. 276, p. 966.

MANUSCRIPT ACCEPTED BY THE SOCIETY 19 APRIL 2005

Printed in the USA

Geological Society of America
Special Paper 395
2005

Reconstruction of subseasonal environmental conditions using bivalve mollusk shells—A graphical model

Bernd R. Schöne
Institute for Geology and Paleontology, University of Frankfurt, Senckenberganlage 32-34, 60325 Frankfurt-Main, Germany

Jon Hickson
School of Education, Health and Sciences, University of Derby, Kedleston Road, Derby, DE22 1GB, United Kingdom

Wolfgang Oschmann
Institute for Geology and Paleontology, University of Frankfurt, Senckenberganlage 32-34, 60325 Frankfurt-Main, Germany

ABSTRACT

This paper presents a graphical model for high-resolution qualitative reconstructions of environmental conditions (temperature, food availability, and fresh-water influx) from bivalve mollusk shells. The growth rate–temperature model (GRT model) is based on the observation that temperature is the most important control of shell growth as shown by shells of *Phacosoma japonicum* (Reeve). A highly significant correlation (p <0.0001) exists between daily shell growth rate of this species and water temperature. Nonparametric bivariate density fields (NBDF) describe the area in which 95% of the growth rate–temperature data are plotted (observed temperatures or temperatures reconstructed from oxygen isotope ratios of the shell). Growth rate–temperature values deviating from the NBDF indicate significant variations in food availability and/or fresh-water influx (salinity changes).

The model is applicable to Recent and fossil bivalve mollusk shells. Its strength lies in the combination of sclerochronological (growth rate analysis) and stable isotope analyses. The integration of other environmental proxies (trace and minor elements and elemental ratios) archived in bivalve shells could further improve the GRT model and allow for high-resolution quantitative environmental reconstructions.

Keywords: bivalve mollusk shell, sclerochronology, oxygen isotope, modeling, temperature, food, fresh water.

INTRODUCTION

Bivalve mollusk shells provide an extremely valuable source material for high-resolution (annual, fortnightly, circadian, ultradian) paleoenvironmental reconstructions. Bivalves grow by periodic accretion of skeletal carbonate (Rao, 1954; Morton, 1970; Richardson et al., 1988) and record environmental information during their growth (e.g., Kennish and Olsson, 1975; Turekian et al., 1982; Williams et al., 1982). Periodic accretion of skeletal material partitions the hard parts into time intervals of near equal duration, e.g., in annual (Barker, 1964), fortnightly (Evans, 1972; Pannella, 1976), and daily growth increments (Evans, 1972; Clark, 1975; Schöne et al., 2002a). Growth increments can allow exact calendar dates to be assigned to each portion of the shells (Pannella and MacClintock, 1968; Koike, 1980; Goodwin et al., 2001). Bivalve shells are excellently suited for multiproxy environmental reconstructions as they preserve environmental information in various ways, i.e., chemical properties of their growth increments (e.g., Wefer and Berger, 1991; Mutvei et al., 1994; Goodwin et al., 2001) and in variable growth rates (Jones

Schöne, B.R., Hickson, J., and Oschmann, W., 2005, Reconstruction of subseasonal environmental conditions using bivalve mollusk shells—A graphical model, *in* Mora, G., and Surge, D., eds., Isotopic and elemental tracers of Cenozoic climate change: Geological Society of America Special Paper 395, p. 21–31, doi: 10.1130/2005.2395(03).

et al., 1989; Marchitto et al., 2000; Schöne, 2003). The rate at which shells grow is controlled by physiological processes such as aging (e.g., Hall et al., 1974; Berta, 1976) and environmental factors, primarily ambient water temperature (Henderson, 1929; Kennish and Olsson, 1975) and food availability (Ansell, 1968; Sato, 1997).

Seasonal and subseasonal environmental data can significantly improve models of past climates and hence put climate forecasting on a firmer foundation (Manabe and Stouffer, 1988; Kumar et al., 1996; Derome et al., 2001; Klein Tank and Können, 2003; Kiktev et al., 2003; Klein Tank et al., 2005). Until now, the few existing models capable of reconstructing daily environmental conditions from bivalve mollusks focus solely on water temperatures (Schöne et al., 2002b). No integrated multiproxy approach exists that studies the full range of environmental information contained within bivalve shells. This goal requires an integrated approach employing both sclerochronology (growth rate analysis) and shell geochemistry. Similar studies have been conducted using corals for more than two decades (e.g., Gagan et al., 1994; Watanabe et al., 2001; Marshall and McCulloch, 2002).

In the present study wc present a graphical model for subseasonal reconstructions of temperature, food availability, and fresh-water influx: the growth rate–temperature (GRT) model. It is based on the observation that shell growth of the shallow-marine veneroid bivalve *Phacosoma japonicum* is mainly controlled by water temperature (Tanabe, 1988; Schöne et al., 2003) but is influenced by additional factors. We test the model with a shell collected alive and measured environmental (temperature, salinity) data and demonstrate its applicability to fossil specimens of *P. japonicum* from the middle and late Holocene. This model is versatile as it requires only slight modifications to reconstruct environmental conditions using other bivalve species.

MATERIAL AND METHODS

Five previously studied specimens of *P. japonicum* were chosen from specimens collected alive from shallow-marine settings at three different localities around Japan (Hakodate Bay, Tokyo Bay, and Seto Inland Sea; Fig. 1; HB1-A10, collected alive during 1987; SI6-A2 and SI7-A2, 1986; TB1-A2 and TB1-A11, 1989). Well-preserved fossil specimens of the same species were recovered from middle and late Holocene intertidal deposits of the Miura Peninsula, Tokyo Bay (Fig. 1; YCM-GP835-32-2, 5270–4800 cal yr B.P.; YCM-GP848-10-2, 6120–5590 cal yr B.P.), collected and provided by Yasumitsu Kanie, former curator of the Yokosuka City Museum. Calibrated radiocarbon ages (cal B.P.) were calculated using CALIB 4.3 (Stuiver and Reimer, 1993; Stuiver et al., 2000). Because no information is available on marine reservoir effects for the exact locality where the shells were collected, published data from the Northwestern Pacific were averaged (average $\Delta R = 7.6 \pm 41.4$; Southon et al., 2002; Konishi et al., 1982; Kuzmin et al., 2001; references for calibration data sets: Stuiver et al., 1998a, 1998b). ^{14}C dating was performed on shells found in the same beds as the samples studied here. The excellent preservation of the specimens precludes postmortem transport. Although time-averaging in nearshore environments does occur (Flessa and Kowalewski, 1994), radiometric ages provide an estimate of the mean ages of shells selected at random from each accumulation.

Lunar daily increment widths were determined from digital images of etched and Au-coated cross sections of the shells using Scion/NIH version 4.02b image analysis software (http://www.scioncorp.com, last visited on January 31, 2005). Daily growth rate was plotted versus temperature. Linear fits and the nonparametric bivariate density (95% levels) were calculated using the

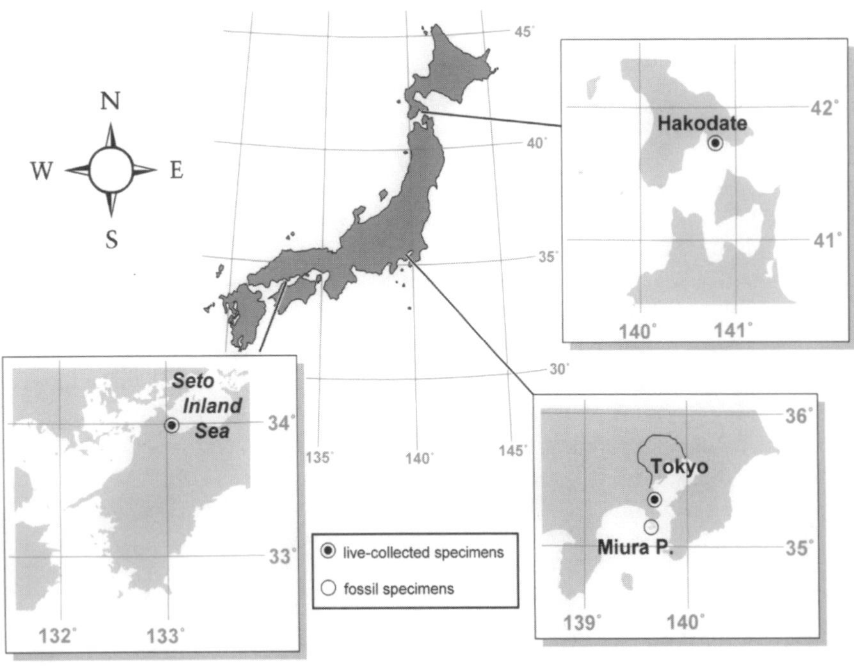

Figure 1. Map showing sample localities around shallow-marine settings of the Japanese coast. Recent shells: Hakodate Bay (HB), 41°49'N, 140°42'E; Tokyo Bay (TB), 35°20'N, 139°42'E; Seto Inland Sea (SI), 33°55'N, 133°03'E. Fossil shells: Miura Peninsula.

statistical software analysis package JMP In 3 by SAS Institute Inc. The nonparametric bivariate density estimates are designed for the situation when it is hard to see patterns in a scatterplot. The bivariate density estimation models a contour line that has 95% of the growth rate–temperature pairs below it.

A total of 42 samples of shell carbonate (each weighing 40–60 µg) were processed for oxygen isotope analysis using a Finnigan MAT 252 mass spectrometer equipped with a Kiel III automated sampling device. Isotope results are reported relative to Vienna Peedee belemnite (VPDB). Temperatures were calculated from oxygen isotope ratios of the shells ($\delta^{18}O_{aragonite}$) using the modified Grossman and Ku (1986) function:

$$T_{\delta^{18}O}\ (°C) = 20.6 - 4.34\ [\delta^{18}O_{aragonite} - (\delta^{18}O_{ocean} - 0.2)]. \quad (1)$$

Thus, assuming no change in the $\delta^{18}O$ of the seawater ($\delta^{18}O_{ocean}$), a one per mil shift in shell $\delta^{18}O_{aragonite}$ reflects a temperature change of the ambient seawater of 4.34 °C. A $\delta^{18}O_{ocean}$ value of 0‰ standard mean ocean water (SMOW) for Recent shells and specimens from Holocene deposits was assumed. Sea level changes since the middle Holocene were only minor (Matsushima, 1984) and had no significant effect on the $\delta^{18}O_{ocean}$ values (Fairbanks, 1989).

Water temperature and salinity data were provided by marine oceanographic stations (Japan Meteorological Agency). Monthly precipitation data (University of Delaware satellite measurements) were downloaded from the National Oceanic and Atmospheric Administration-Cooperative Institute for Research in Environmental Sciences (NOAA-CIRES) Climate Diagnostics Center (http://www.cdc.noaa.gov, last visited on January 31, 2005). Fluctuations in salinity provide a good indicator of fresh-water influx into the bivalve habitat. Phytoplankton pigment concentration data were obtained from the National Aeronautics and Space Administration (Nimbus-7 Coastal Zone Color Scanner). Daily data were interpolated from the environmental measurements through the application of regression analysis (best fits) using the freeware CurveExpert ver. 1.37 (http://curveexpert.webhop.biz/, last visited on January 31, 2005). For comparison of the data sets, the discrepancy between the length of a solar (24 h; daily environmental data) and a lunar day (~24.8 h; daily growth increments), roughly 50 min, was corrected by deleting every 30th value from the daily environmental data sets.

RECENT SHELLS OF *PHACOSOMA JAPONICUM*

Temperature-Controlled Shell Growth

Recently, a detailed study was published on how environmental parameters control shell growth rates of *P. japonicum* (Schöne et al., 2003). The most important finding was that water temperature and the amount and quality of food are the main controls of shell growth (see also Tanabe, 1988; Sato, 1999).

A positive relationship was found between temperature and shell growth rate. Shell growth only occurs between 14.2 °C and 16.8 °C and ~29 °C (shutdown temperatures) and is fastest between 24.6 °C and 27.2 °C (optimal growth temperature) during hot summer. During the first half of the year (before hot summer), water temperature (*T*) and lunar daily growth increment (*LDGI*) widths are linearly correlated with each other (Fig. 2). Irrespective of the geographic position, *P. japonicum* grows at the same rate at similar temperatures if food is neither sharply reduced nor available in plentiful supply (Schöne et al., 2004).

The relationship between growth rate and temperature varies with ontogenetic age. Specimens of the same age exhibit similar slopes. However, slopes are steeper in younger specimens than in older ones (Fig. 2; see also Schöne et al., 2003). A 1 °C temperature change results in an *LDGI* change of ~16–19 µm in two- and three-year-old specimens, but only ~6 µm in four-year-old specimens. This finding confirms previous observations by Schöne et al. (2003, 2004).

Previous results also indicate that variable fresh-water influx, and hence salinity changes, do not significantly affect shell growth rates, but shift the $\delta^{18}O_{aragonite}$ ratios of the shells to more negative values. $\delta^{18}O_{aragonite}$ values as negative as –2.55‰ occur during the rainy season in June and July (Schöne et al., 2003). These values correspond to a temperature of 33.33 °C, which is ~4.3 °C higher than those measured in the field (29 °C). Thus, the annual oxygen isotope profiles of the shells reflect both the temperature cycle and the varying amounts of fresh water of more negative $\delta^{18}O$ added to the seawater by precipitation and rivers.

Shell Growth Controlled by Temperature and Food Availability

A previous study (Schöne et al., 2003) demonstrated that specimens alive during eutrophic conditions grow at much higher rates, and, consequently, the lines of correlation between *LDGI* and *T* are steeper than in other specimens. Figure 3 exemplifies this finding. The *LDGI* width of the three-year-old specimen TB1-A11 from Tokyo Bay (Table 1) increased quickly in late March 1985 and reached a maximum of ~200 µm per day at temperatures of 23 °C in July/August (Fig. 3A). During spring and early summer the phytoplankton concentration reached very high levels (Fig. 3E), which supported fast shell growth rates (Fig. 3A). After the hottest part of the year, growth rates decreased and the shell stopped growing when temperatures fell below the lower shutdown temperature in December (Figs. 3A and 3C). Shell growth in specimen TB1-A11 is not affected by fresh-water influx or reduced salinity occurring during the summer monsoon or typhoon storms (Figs. 3A and 3E). However, the influx of fresh water (more negative $\delta^{18}O$) drives the $\delta^{18}O_{aragonite}$ ratios to more negative values (Fig. 3B).

FOSSIL SHELLS OF *PHACOSOMA JAPONICUM*

The specimens used here were analyzed in detail by Schöne et al. (2004). Fossil specimens of *P. japonicum* grew at similar rates as modern specimens; each lunar day one LDGI was

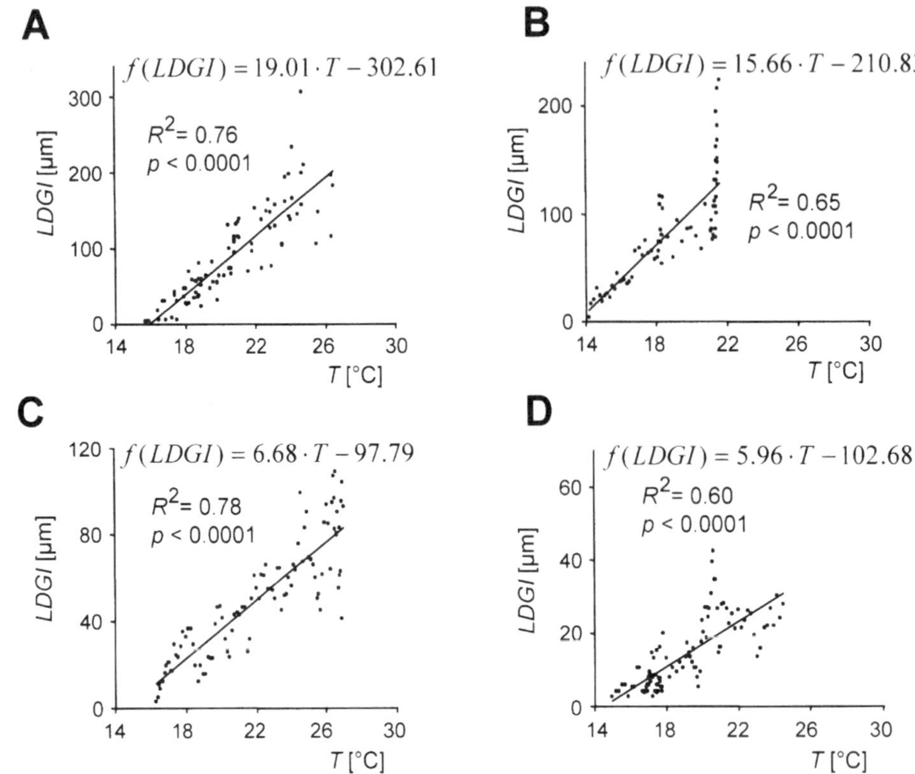

Figure 2. *Phacosoma japonicum*. (A) Seto Inland Sea, specimen SI7-A2, age two, 1985; (B) Hakodate Day, specimen HB1-A10, age three, 1976; (C) Seto Inland Sea, specimen SI6-A2, age four, 1984; (D) Tokyo Bay, specimen TB1-A2, age four, 1986. During the first half of the year (before summer) lunar daily growth rates (*LDGI*) and water temperature (*T*) are highly significantly correlated. Food is neither overabundant nor restricted during that time at these localities. At similar temperatures shells of the same age class grow at similar rates. A temperature increase of 1 °C results in an *LDGI* increase of 16–19 µm in two- and three-year-old specimens (Figs. 2A and 2B) and 6 µm in four-year-old specimens (Figs. 2C and 2D). The shells presented here were studied in detail by Schöne et al. (2003) and provide the basis for the growth rate–temperature (GRT) model (Fig. 5).

laid down (Schöne et al., 2004). Intra-annual dating was based on comparison with growth patterns of recent specimens and $\delta^{18}O_{aragonite}$ profiles. For example, shell portions with the most negative $\delta^{18}O_{aragonite}$ values were probably deposited during late August, whereas portions with sharp $\delta^{18}O_{aragonite}$ peaks are related to the monsoonal rainfall in June and July (for details see Schöne et al., 2004).

Specimen YCM-GP835-32-2 (late Holocene; Fig. 4A) grew from about late April through October and at maximum rates in early August and mid-September. An intra-annual growth break was detected about halfway between the winter breaks, i.e., in about late August. Maximum growth rates were close to 120 µm per day during August. $\delta^{18}O_{aragonite}$ values range between −1.39‰ and +0.51‰.

In contrast, the intra-annual growth pattern of specimen YCM-GP848-10-2 (middle Holocene; Fig. 4B) differs significantly from modern shells. The shell started growing in about early February and reached maximum growth rates of nearly 200 µm per day in late April. A growth break of uncertain duration occurred at the beginning of July. The shell commenced growing for a month or so in late fall. Both the *LDGI* and $\delta^{18}O_{aragonite}$ curves appear strongly asymmetric because shell growth ceased during major parts of the second half of the year. The most negative $\delta^{18}O_{aragonite}$ values are as low as −2.55‰.

GRT MODEL: TEMPERATURE-CONTROLLED SHELL GROWTH

Based on the shell growth–temperature relationship, we have established a growth rate–temperature model (GRT model; Fig. 5) with temperature *T* as the independent variable (*x*-axis) and *LDGI* as the dependent variable (*y*-axis). Our model uses only *T* and *LDGI* data of the first half of the year (= no food limitation) of the following specimens: HB1-A10, SI7-A2, SI6-A2, and TB1-A2. Food is neither limited nor overabundant before hot summer at these localities. The linear regression analysis [GRT_{corr}-line]

$$f(LDGI) = 16.26 \cdot T - 235.23 \qquad (2)$$

of observed temperature *T* and *LDGI* values of the two ontogenetically younger specimens (HB1-A10 and SI7-A2) returns

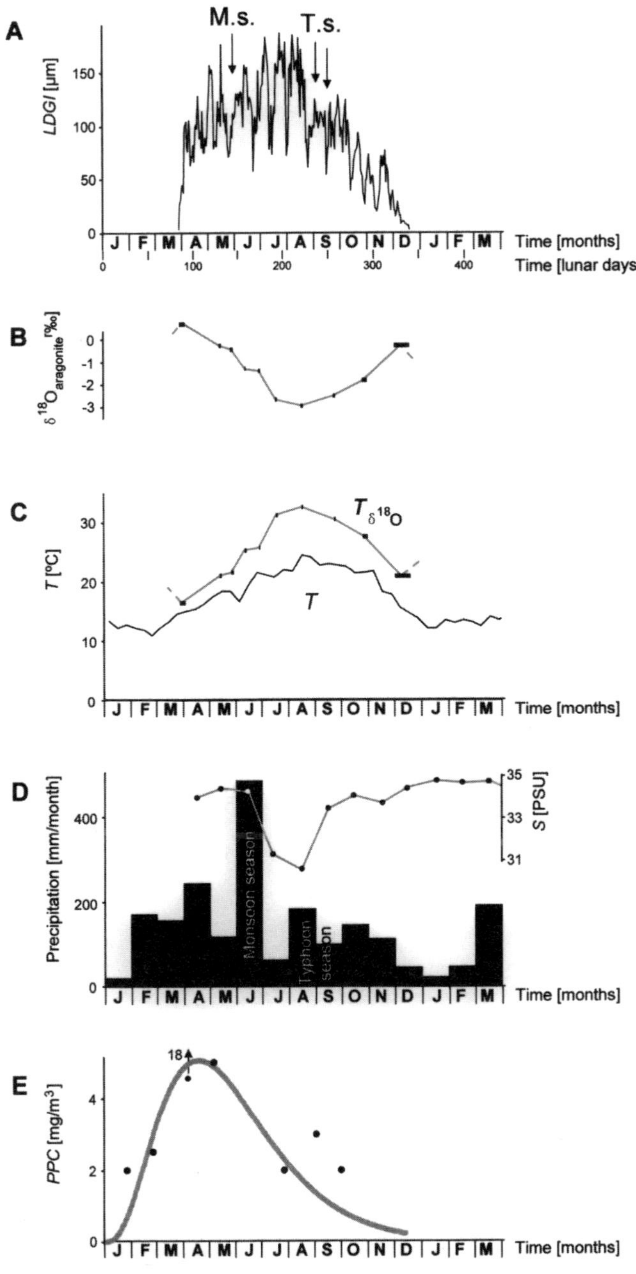

Figure 3. *Phacosoma japonicum*. Comparison of lunar daily growth rates (*LDGI*; Fig. 3A), shell $\delta^{18}O$ values ($\delta^{18}O_{aragonite}$; B+C), and environmental parameters (Figs. 3C–3E) of specimen TB1-A11 alive at Tokyo Bay during 1985. Growth rate and the length of the growing season are mainly controlled by temperature (Figs. 3A and 3C). Observed water temperature (*T*) is plotted with temperature calculated from $\delta^{18}O_{aragonite}$ ($T_{\delta^{18}O}$; Fig. 3C). Disagreement between calculated and observed *T* (Fig. 3C) may be due to the assumption of 0‰ for the seawater oxygen isotope value. High precipitation rates [salinity changes during the monsoon (M.s.) or typhoon season (T.s.)] reduce growth rates only slightly (A+D). The stable isotope composition of the shell (Fig. 3B) is, however, shifted to more negative values during the rainy seasons (Fig. 3D), leading to higher calculated *T* (Fig. 3C). Low nutrient availability after the hot summer (Fig. 3E) reduces *LDGI* (Fig. 3A), *PPC*—phytoplankton pigment concentration. Precipitation data were obtained from NOAA-CIRES Climate Diagnostics Center, Boulder, Colorado, United States (www.cdc.noaa.gov), temperature and salinity data from the Japan Meteorological Agency, and phytoplankton concentration data from NASA (Nimbus-7 CZCS project).

highly significant (p <0.0001) R^2 values of 0.67 (Fig. 5A). The older individuals (TB1-A11 and SI6-A2) grew more slowly and, accordingly, the GRT_{corr}-line is flatter:

$$f(LDGI) = 6.58 \cdot T - 106.51 \quad (3)$$

(p <0.0001, R^2 = 0.67; Fig. 5B). On average, 67% of the variability in daily growth rates can be explained by temperature changes.

In addition to the GRT_{corr}-lines, we depicted the 95% nonparametric bivariate density of the *LDGI-T*-pairs (Fig. 5).

Application of the GRT Model

Food Availability

Hypothesis I. If the water temperature during shell formation is known, the GRT model can provide information on the relative food availability (Fig. 6). *LDGI-T* data plotting outside the NBDF can reflect changes in the food supply. Data plotting to the left of the NBDF indicate increased food availability and faster growth rates than predicted. Data plotting to the right of the NBDF indicate food scarcity and slower growth rates than predicted.

Test of Hypothesis I. We tested this hypothesis using the shell portion formed during ontogenetic year three of specimen TB1-A11 (Fig. 3A). Some of the *LDGI-T* data of this specimen clearly plot outside the NBDF (Fig. 6B). We removed the temperature control from the growth data by subtracting the predicted *LDGI* values (as given by the GRT_{corr}-lines) from the measured *LDGI* values. The residuals ($LDGI_T$) were then plotted against the phytoplankton pigment concentration data (*PPC;* Fig. 6C). We found that the $LDGI_T$ and *PPC* values are highly significantly correlated with each other (R^2 = 0.69; p <0.0001). This strongly suggests that higher food levels can promote shell growth.

We can also add a calendar to the $LDGI_T$ data and demonstrate when during the growing season food levels were higher or lower (for details see Schöne et al., 2004). Then the $LDGI_T$ and the *PPC* values were plotted versus time (Fig. 6D). Specimen TB1-A11 grew much faster than predicted from the GRT model during spring and early summer, but slightly less than predicted during hot summer and fall.

Finally, we show how the *PPC* levels control the position of the *LDGI-T* data in the GRT model. For this purpose, we plotted the residuals of the $LDGI_T$–*PPC* plot, $LDGI_{T,PPC}$, against *T* in the GRT model. As expected the $LDGI_{T,PPC}$–*T* values plot farther off the NBDF (Fig. 6E).

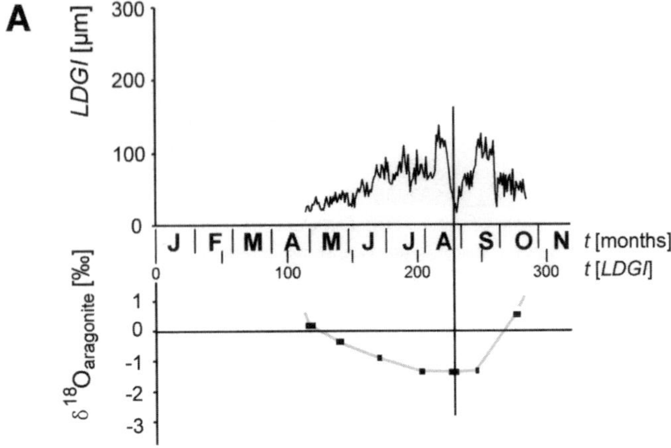

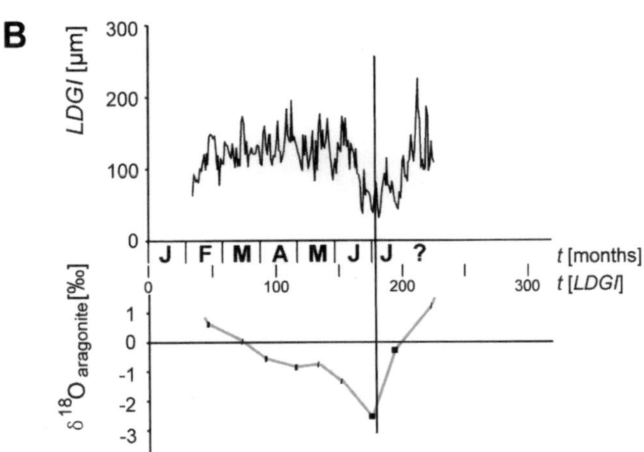

Figure 4. *Phacosoma japonicum.* Lunar daily growth rates (*LDGI*) and temperatures derived from $\delta^{18}O_{aragonite}$ ($T_{\delta^{18}O}$) are depicted for two fossil shells. Specimen YCM-GP835-32-2 (age four; Fig. 4A) came from the late Holocene. Specimens YCM-GP848-10-2 (age two; Fig. 4B) lived during the mid-Holocene Climate Optimum. Intra-annual dating was done by comparison with $\delta^{18}O_{aragonite}$ curves and *LDGI* patterns observed in modern shells. For instance, sharp $\delta^{18}O_{aragonite}$ peaks indicate shells portions deposited during the monsoon season (June and July). Growth breaks occurred during the hottest season in both shells (solid vertical line). The late Holocene shell grew almost year round and shows only a short summer growth break, possibly a spawning break. The shell from the middle Holocene, however, did not grow for major parts of the second half of the year. Calcification stopped at around early July, perhaps due to hot temperatures and food deprivation. Shell oxygen isotope ratios of shell portions deposited during June and July suggest that the climate was very wet.

Fresh-water Influx

Hypothesis II. The GRT model can also graphically detail information on variable fresh-water influx if pairs of $LDGI_{T,PPC}$ and $\delta^{18}O_{aragonite}$–derived temperature ($T_{\delta^{18}O}$) data are plotted on top of the GRT model (Fig. 7A). $LDGI_{T,PPC}$–$T_{\delta^{18}O}$ data plotting left of the NBDF indicate evaporation. The influx of large amounts of fresh water enriched in ^{16}O causes the opposite: $\delta^{18}O_{aragonite}$ values become more negative and, consequently, $T_{\delta^{18}O}$ values increase. The farther right the *LDGI*–$T_{\delta^{18}O}$ data plot to the right of the NBDF, the higher the fresh-water influx and the lower the salinity.

Test of Hypothesis II. Hypothesis II was tested on specimen TB1-A11. As depicted in Figure 7B, $LDGI_{T,PPC}$–$T_{\delta^{18}O}$ data plot mainly to the right of the NBDF and are shifted to the right (along the *x*-axis) relative to the $LDGI_{T,PPC}$–*T* data (in gray).

We subtracted the measured temperatures from the $\delta^{18}O_{aragonite}$–derived temperatures ($T_{diff} = T_{\delta^{18}O} - T$). Then we plotted T_{diff} versus salinity S (as a proxy for fresh-water influx; Fig. 7C). T_{diff} is significantly correlated with *S* ($R^2 = 0.51, p = 0.034$).

$$f(S) = -0.34 \cdot T_{diff} + 35.31. \quad (4)$$

Hence, T_{diff} is a measure of fresh-water influx. $LDGI_{T,PPC}$–$T_{\delta^{18}O}$ data plotting to the right of the NBDF of specimen TB1-A11 (Fig. 7B) reflect the addition of fresh water to the seawater by precipitation and rivers.

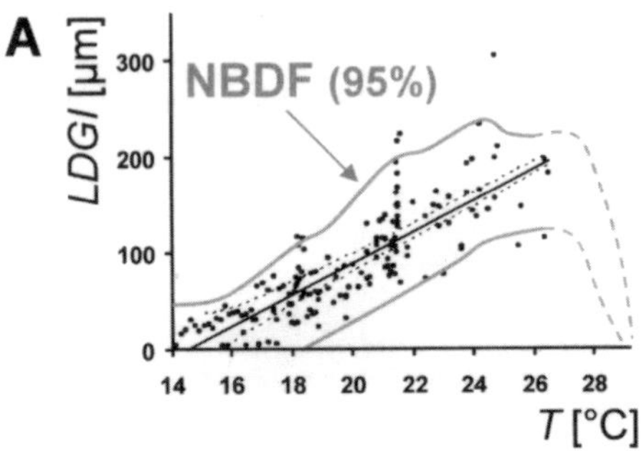

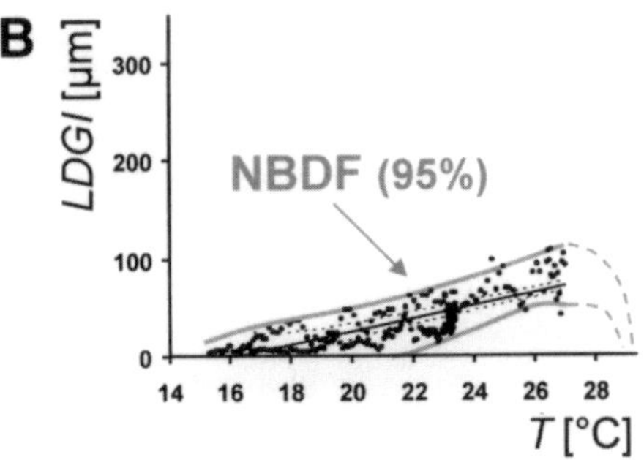

Figure 5. Growth rate–temperature (GRT) model for younger (ages two and three; Fig. 5A) and older (age four; Fig. 5B) specimens of *Phacosoma japonicum.* Figure 5A uses lunar daily growth rates–temperature (*LDGI–T*) data for the first half of the year for specimens SI7-A2 (Fig. 2A) and HB1-A10 (Fig. 2B), and Figure 5B uses the data for specimens SI6-A2 (Fig. 2C) and TB1-A2 (Fig. 2D). 95% nonparametric bivariate density fields (NBDF) indicate the area in which 95% of the *LDGI–T* data plot. *LDGI–T* data of other shells plotting outside the NBDF reflect changes in temperature or food availability; *LDGI*–$T_{\delta^{18}O}$ pairs deviating from the NBDF, however, reflect changes in salinity (oxygen isotope composition of the seawater by fresh-water influx of different $\delta^{18}O$).

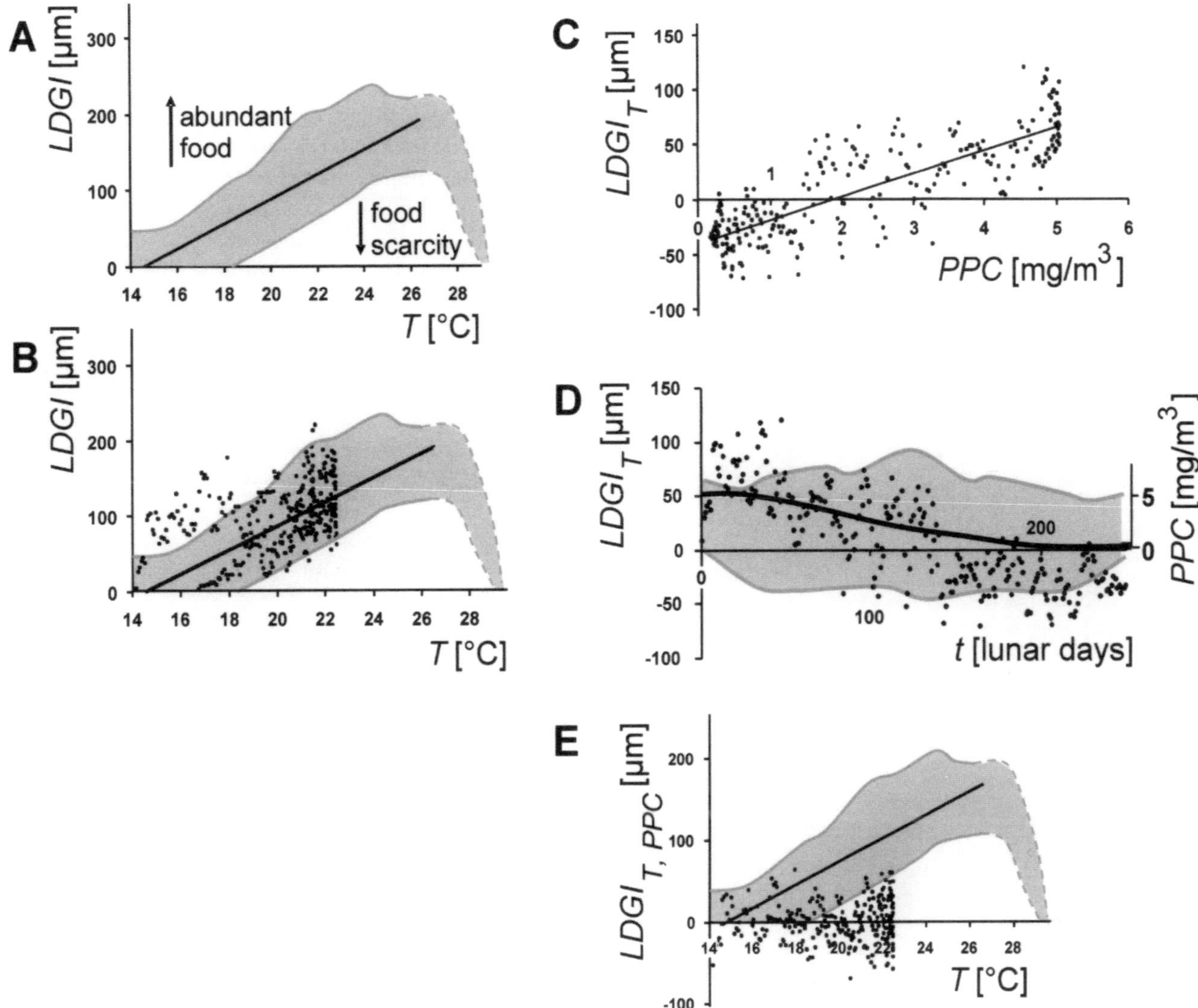

Figure 6. *Phacosoma japonicum.* Reconstruction of food availability using the growth rate–temperature (GRT) model. (A) Hypothesis I: lunar daily growth rates–temperature (*LDGI–T*) pairs plotting to the right of the nonparametric bivariate density fields (NBDF) of the GRT model (Fig. 5) indicate abundant food supply, and data plotting to the left of the NBDF reflect growth during food deprivation. (B–E) Test of hypothesis I. (B) Some *LDGI–T* data of specimen TB1-A11 plot to the left of the NBDF and reflect growth during limited food availability (see Fig. 3). (C) The temperature control was removed from the *LDGI* widths and the residuals ($LDGI_T$) plotted versus phytoplankton pigment concentration (*PPC*). The highly significant positive linear relation suggests that higher food levels promote growth rates. (D) When plotted versus time, $LDGI_T$ and *PPC* data show when during the year food was limited or abundantly available and how shell growth was affected. (E) If the temperature and the food availability control on shell growth is removed, the residuals ($LDGI_{T,PPC}$) show only little variation eventually related to physiology.

$T_{\delta^{18}O}$ values can provide minimum estimates of changes in salinity (Fig. 7D). $T_{\delta^{18}O}$ exceeding the upper growth temperature of the bivalves, i.e., ~29 °C, reflect the sole contribution of fresh-water influx to the seawater. As suggested by the slopes of the T_{diff}-*S* linear fit (equation number 4), a T_{diff} change of 1 °C corresponds to a change of salinity of 0.34 PSU. Hence, maximum $\delta^{18}O_{aragonite}$–derived temperatures of 32.45 °C recorded in specimen TB1-A11 correspond to a minimum reduction of seawater salinity by 1.18 PSU. Salinity values could be used to estimate minimum amounts of fresh-water influx if the $\delta^{18}O$ of the fresh water are known.

Fresh-water Influx and Food Availability

If measured temperatures are not available, for instance in fossil shells, $\delta^{18}O_{aragonite}$ values can be used as temperature proxies (Epstein et al., 1953). However, $\delta^{18}O_{aragonite}$ values reflect both temperature ($T_{\delta^{18}O}$) and the $\delta^{18}O_{ocean}$ ratios. Variable fresh-water influx of more negative $\delta^{18}O$ can shift the $\delta^{18}O_{ocean}$ toward lighter values. This results in overestimation of the actual water temperatures. Hence, *LDGI*–$T_{\delta^{18}O}$ plots cannot detail particular information on a single parameter (temperature and fresh-water influx or salinity).

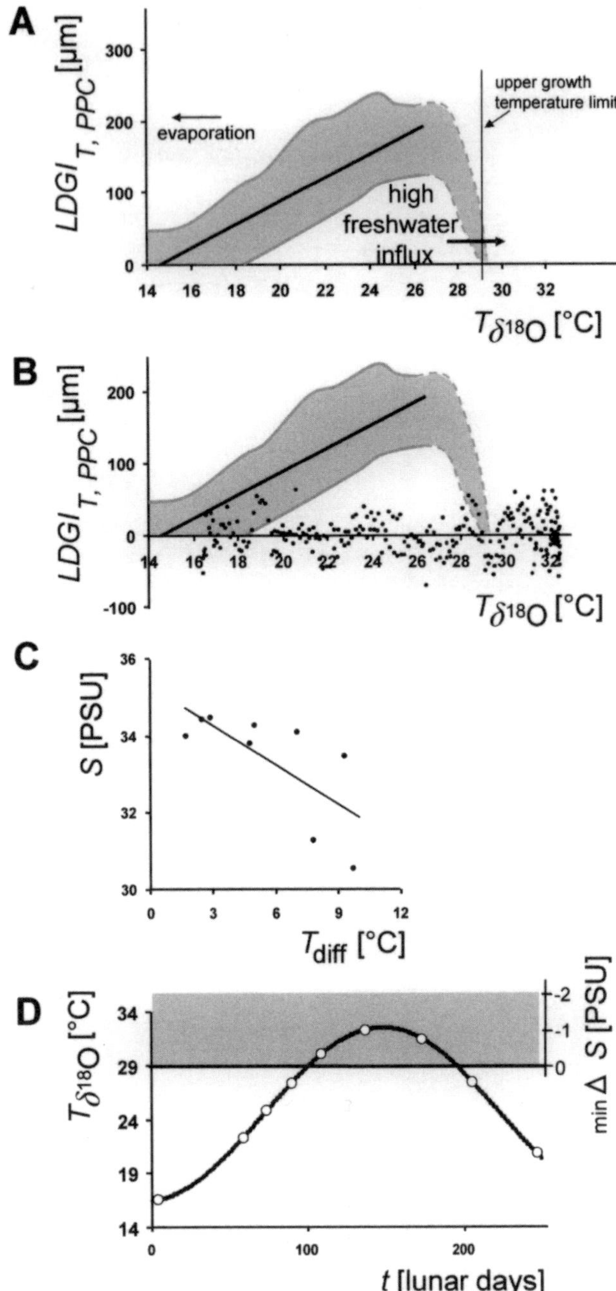

Figure 7. *Phacosoma japonicum.* Reconstruction of fresh-water influx to seawater using the growth rate–temperature (GRT) model. (A) Hypothesis II: $LDGI_{T,PPC}$–$T_{\delta^{18}O}$ (*LDGI*—lunar daily growth rates; *T*—temperature) plots deviating from the nonparametric bivariate density fields (NBDF) refer to changes of the fresh-water influx. Values plotting to the left of the NBDF indicate evaporation, values plotting to the right of the NBDF indicate high fresh-water influx (of lighter $\delta^{18}O$) and hence salinity reduction; *PPC*—phytoplankton pigment concentration. (B–D) Test of hypothesis II. (B) $LDGI_{T,PPC}$–$T_{\delta^{18}O}$ data of specimen TB1-A11 (Figs. 3 and 6E) clearly plot far to the right of the NBDF. (C) Observed temperatures (*T*; Fig. 3C) were subtracted from $T_{\delta^{18}O}$. The results (T_{diff}) plotted versus salinity as a measure of fresh-water influx (compare Fig. 3D) show a significant negative correlation. (D) Reconstructed temperatures ($T_{\delta^{18}O}$) exceeding the upper growth temperature of *P. japonicum* (29 °C) translate into minimum salinity changes. A temperature shift of 1 °C produces a change in $\delta^{18}O$ equivalent to a change in salinity by 0.34 PSU.

When plotted on top of the GRT model, *LDGI*–$T_{\delta^{18}O}$ pairs plotting to the left of the GRT_{corr}–lines indicate increased food availability and/or evaporation (Fig. 8A). However, evaporation does not play a major role in the region studied here. The influx of large amounts of fresh water enriched in ^{16}O causes the opposite: $\delta^{18}O_{aragonite}$ values become more negative and, consequently, $T_{\delta^{18}O}$ values increase. In addition to fresh-water influx, values plotting to the right of the NBDF can reflect the effect of food scarcity on shell growth rates. The farther right the *LDGI*–$T_{\delta^{18}O}$ data plot to the right of the NBDF, the higher the fresh-water influx by rivers and precipitation (and the lower the salinity).

Paleoenvironmental Reconstructions

The GRT model for fossil shells depicted in Figure 8A provides a useful tool for paleoenvironmental reconstructions from $\delta^{18}O_{aragonite}$ and *LDGI* data. In the following we show how two fossil intertidal shells of *P. japonicum* can reconstruct subseasonal environmental conditions during the middle and late Holocene.

LDGI–$T_{\delta^{18}O}$ data of specimen YCM-GP835-32-2 (late Holocene; Figs. 8B and 8C) plot within the NBDF or slightly to the left of the NBDF, suggesting that, first, $T_{\delta^{18}O}$ values closely resemble actual water temperatures (almost 26 °C during hot summer), and second, that food supply was slightly enriched. However, precipitation rates remained very low. This growth pattern is typical for specimens alive today at Hakodate Bay.

In contrast, the GRT plot of specimen YCM-GP848-10-2 (middle Holocene; Figs. 8D and 8E) is more similar to specimens from Tokyo Bay and Seto Inland Sea. Prior to the rainy season *LDGI*–$T_{\delta^{18}O}$ data plot within the NBDF and suggest that $T_{\delta^{18}O}$ data closely resemble actual water temperatures, for instance 26 °C during April and May. This is ~3 °C higher than today at this period of the year (see Schöne et al., 2003). While *LDGI*–$T_{\delta^{18}O}$ data of shell portions deposited during February plot to the left of the NBDF (Fig. 8D), values of June and July plot far off on the right side of the NBDF (Fig. 8E). Temperatures during June and July derived from $\delta^{18}O_{aragonite}$ exceed the upper shutdown temperature by almost 2 °C. This indicates abundant food supply during February and most probably heavy monsoonal precipitation during June and July. Salinity was at least 0.7 PSU below normal levels. Higher temperatures and food scarcity in August probably caused the growth break. *LDGI*–$T_{\delta^{18}O}$ data of shell material produced after the summer break plots to the left of the NBDF and is best explained by phytoplankton blooms in late fall.

Results on environmental conditions obtained from shells of *P. japonicum* using the GRT model suggest that the climate during the middle Holocene was warm and wet, whereas drier and cooler conditions prevailed during the late Holocene. This compares well to climate reconstructions based on other proxy data. The middle Holocene is often regarded as the "Holocene Climatic Optimum" (MacCracken et al., 1990), a period of warm and wet climate favorable for the rise of civilizations and beneficial for a high biodiversity. Japan and adjacent areas expe-

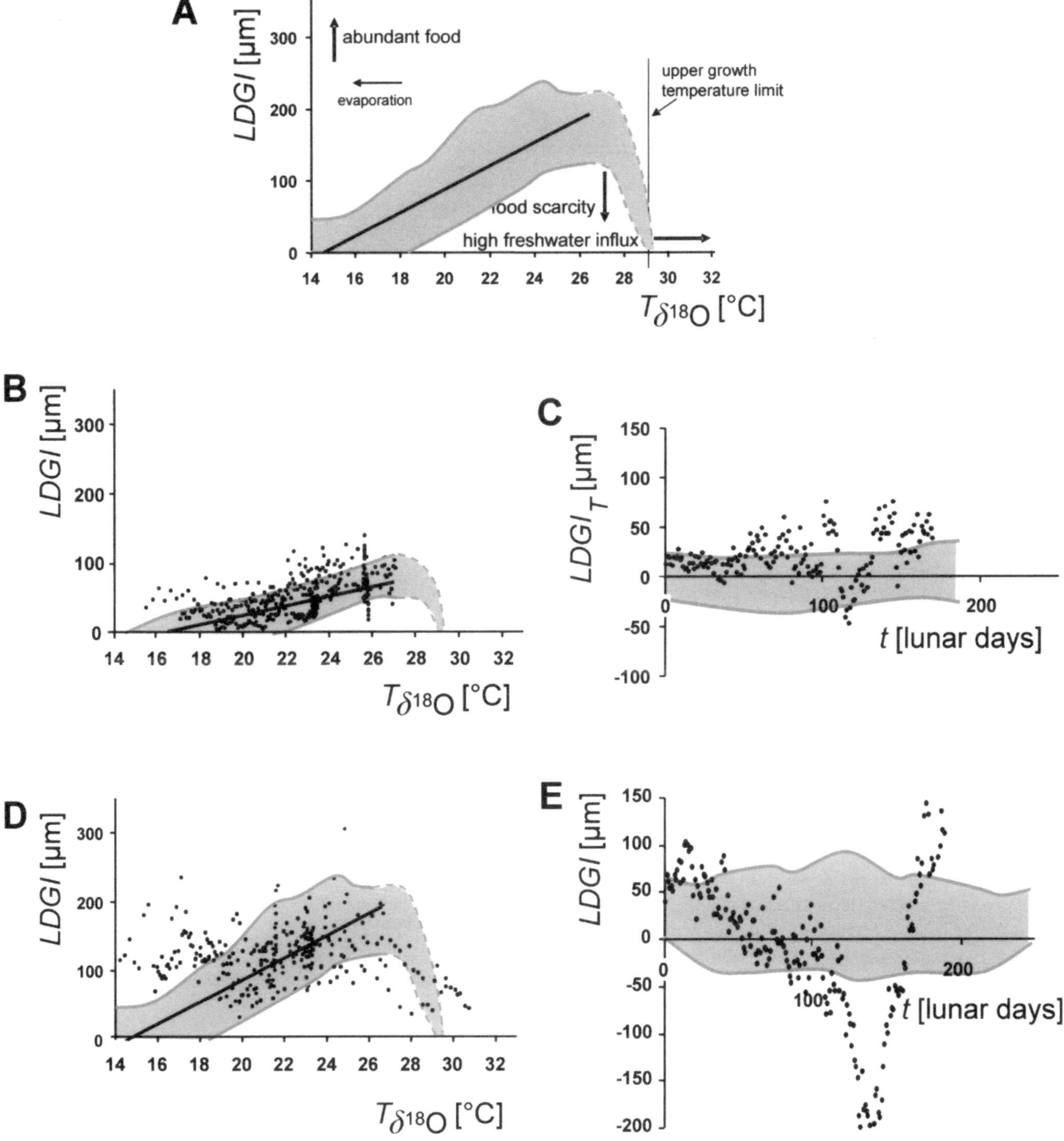

Figure 8. *Phacosoma japonicum.* Reconstruction of fresh-water influx to the seawater and food availability using the growth rate–temperature (GRT) model. Application of the GRT model for fossil shells. (A) *LDGI*–$T_{\delta^{18}O}$ (*LDGI*—lunar daily growth rate; *T*—temperature) pairs plotting to the left of the nonparametric bivariate density fields (NBDF) are related to low food availability. Evaporation did not play a major role in study area. Data plotting to the right of the NBDF are equivocal as they can indicate either enhanced fresh-water influx or food scarcity. (B) *LDGI*–$T_{\delta^{18}O}$ pairs of specimen YCM-GP835-32-2 (Fig. 4A) plot mostly within the NBDF or to the left of the NBDF. (C, E) $LDGI_T$ time series reveal when during the year growth was lower or higher than expected from the GRT model. (D) *LDGI*–$T_{\delta^{18}O}$ pairs of specimen YCM-GP848-10-2 (Fig. 4B) plot far off to both sides of the NBDF.

rienced the "Holocene Climatic Optimum" around 6500–5500 yr B.P. (e.g., Tsukada, 1986; Sakaguchi, 1989; Pflaumann and Jian, 1999; Takahara et al., 2000; Ternois et al., 2000). After that period a cooling trend began around 5500–5000 yr B.P. (Yasuda, 1983; Abram et al., 2001; Grossman, 2001).

STRENGTHS AND LIMITS OF THE GRT MODEL

The GRT model combines sclerochronological and isotope geochemical data to reconstruct temperature, food availability, and precipitation from bivalve mollusk shells. Because shells of many bivalve mollusk species form in close relation to the ambient water temperature, growth rate–temperature plots can inform about changes of environmental parameters such as food availability and fresh-water influx. One of the strengths of the GRT model is its graphical approach at the reconstruction of environmental parameters in recent and fossil environments.

Up to now, however, the model works only reliably for modern shells. It remains difficult to discern between the relative contribution of variable fresh-water influx and food availability in fossil shells. For example, $LDGI–T_{\delta^{18}O}$ pairs of shell portions formed during relatively low temperatures, abundant food supply, and large fresh-water influx can plot within the NBDF. Our model would falsely suggest that the shell formed at higher temperatures. However, in the habitat in which *P. japonicum* grows, large fresh-water inflow occurs only after the monsoon in summer when food availability has already decreased. A multiproxy approach including, for example, Sr/Ca ratios (Beck et al., 1992) could reduce the likelihood of erroneous interpretations.

CONCLUSIONS

GRT models provide an excellent tool to reconstruct recent and fossil environmental conditions from bivalve shells. Growth rate–temperature (measured or reconstructed from $\delta^{18}O_{aragonite}$) data are compared to reference data limited by the 95% NBDF determined for modern shells. In the case of *P. japonicum* from Japan, NBDF separate two growth conditions from each other: increased food availability to the left side of the NBDF and mainly increased precipitation to the right side of the NBDF.

Through the integration of additional independent proxy data the GRT model could evolve into a numerical model and provide quantitative information on paleoenvironmental conditions.

ACKNOWLEDGMENTS

We thank David Dettman (University of Arizona, Tucson, Arizona, United States) for some of the isotope measurements. Yasumitsu Kanie kindly provided fossil shells of *P. japonicum*. Two anonymous reviewers helped improve a former draft of the manuscript. We thank Germán Mora and Donna Surge for guest editing this special issue. We are also indebted to David Rodland for comments on the text. This study has been made possible by a German Research Foundation (DFG) grant within the framework of the Emmy Noether Program (SCHO 793/1).

REFERENCES CITED

Abram, N.J., Webster, J.M., Davies, P.J., and Dullo, W.C., 2001, Biological response of coral reefs to sea surface temperature variation: Evidence from the raised Holocene reefs of Kikai-jima (Ryukyu Islands, Japan): Coral Reefs, v. 20, p. 221–234, doi: 10.1007/s003380100163.

Ansell, A.D., 1968, The rate of growth of the hard clam *Mercenaria mercenaria* (L) throughout the geographic range: Journal du Conseil Permanent International pour l'Exploration de la Mer, v. 31, p. 364–409.

Barker, R.M., 1964, Microtextural variation in pelecypod shells: Malacologia, v. 2, p. 69–86.

Beck, J.W., Edwards, R.L., Ito, E., Taylor, F.W., Recy, J., Rougerie, F., Joannot, P., and Henin, C., 1992, Sea-surface temperature from coral skeletal strontium/calcium ratios: Science, v. 257, p. 644–647.

Berta, A., 1976, An investigation of individual growth and possible age relationships in a population of *Protothaca staminea* (Mollusca: Pelecypoda): PaleoBios, v. 21, p. 1–25.

Clark, G.R., II, 1975, Periodic growth and biological rhythms in experimentally grown bivalves, *in* Rosenberg, G.D., and Runcorn, S.K., eds., Growth Rhythms and the History of the Earth's Rotation: London, Wiley, p. 103–117.

Derome, J., Brunet, G., Plante, A., Gagnon, N., Boer, G.J., Zwiers, F.W., Lambert, S.J., Sheng, J., and Ritchie, H., 2001, Seasonal predictions based on two dynamical models: Atmosphere-Ocean, v. 39, p. 485–501.

Epstein, S., Buchsbaum, R., Lowenstam, H.A., and Urey, H.C., 1953, Revised carbonate-water isotopic temperature scale: Geological Society of America Bulletin, v. 64, p. 1315–1326.

Evans, J.W., 1972, Tidal growth increments in the cockle *Clinocardium nuttalli:* Science, v. 176, p. 416–417.

Fairbanks, R.G., 1989, A 17,000-year glacio-eustatic sea level record: Influence of glacial melting rates on the Younger Dryas event and deep-ocean circulation: Nature, v. 342, p. 637–642, doi: 10.1038/342637a0.

Flessa, K.W., and Kowalewski, M., 1994, Shell survival and time-averaging in nearshore and shelf environments—Estimates from the radiocarbon literature: Lethaia, v. 27, p. 153–165.

Gagan, M.K., Chivas, A.R., and Isdale, P.J., 1994, High-resolution isotopic records from corals using ocean temperature and mass-spawning chronometers: Earth and Planetary Science Letters, v. 121, p. 549–558, doi: 10.1016/0012-821X(94)90090-6.

Goodwin, D.H., Flessa, K.W., Schöne, B.R., and Dettman, D.L., 2001, Cross-calibration of daily growth increments, stable isotope variation, and temperature in the Gulf of California bivalve mollusk *Chione cortezi:* Implications for paleoenvironmental analysis: Palaios, v. 16, p. 387–398.

Grossman, E.L., and Ku, T.-L., 1986, Oxygen and carbon isotope fractionation in biogenic aragonite; temperature effects: Chemical Geology, Isotope Geoscience Section, v. 59, p. 59–74, doi: 10.1016/0168-9622(86)90057-6.

Grossman, M.J., 2001, Large floods and climatic change during the Holocene on the Ara River, Central Japan: Geomorphology, v. 39, p. 21–37, doi: 10.1016/S0169-555X(01)00049-6.

Hall, C.A., Jr., Dollase, W.A., and Corbató, C.E., 1974, Shell growth in *Tivela stultorum* (Mawe, 1823) and *Callista chione* (Linnaeus, 1758) (Bivalvia): Annual periodicity, latitudinal differences, and diminution with age: Palaeogeography, Palaeoclimatology, Palaeoecology, v. 15, p. 33–61, doi: 10.1016/0031-0182(74)90036-4.

Henderson, J.T., 1929, Lethal temperatures of Lamellibranchiata: Contributions to Canadian Biology and Fishery, v. 4, p. 399–411.

Jones, D.S., Arthur, M.A., and Allard, D.J., 1989, Sclerochronological records of temperature and growth from shells of *Mercenaria mercenaria* from Narragansett Bay, Rhode Island: Marine Biology, v. 102, p. 225–234, doi: 10.1007/BF00428284.

Kennish, M.J., and Olsson, R.K., 1975, Effects of thermal discharges on the microstructural growth of *Mercenaria mercenaria:* Environmental Geology (Springer), v. 1, p. 41–64.

Kiktev, D., Sexton, D.M.H., Alexander, L., and Folland, C.K., 2003, Comparison of modeled and observed trends in indices of daily climate extremes: Journal of Climate, v. 16, p. 3560–3571, doi: 10.1175/1520-0442(2003)016<3560:COMAOT>2.0.CO;2.

Klein Tank, A.M.G., and Können, G.P., 2003, Trends in indices of daily temperature and precipitation extremes in Europe, 1946–99: Journal of Climate, v. 16, p. 3665–3680, doi: 10.1175/1520-0442(2003)016<3665:TIIODT>2.0.CO;2.

Klein Tank, A.M.G., Können, G.P., and Selten, F.M., 2005, Signals of anthropogenic influence on European warming as seen in the trend patterns of daily temperature variance: International Journal of Climatology, v. 25, p. 1–16, doi: 10.1002/joc.1087.

Koike, H., 1980, Seasonal dating by growth-line counting of the clam, *Meretrix lusoria:* University Museum, University of Tokyo, Bulletin, v. 18, p. 1–120.

Konishi, K., Tanaka, T., and Sakanoue, M., 1982, Secular variation of radiocarbon concentrations in seawater: Sclerochronological approach: Proceedings of the Fourth International Coral Reef Symposium, v. 1, p. 181–185.

Kumar, A., Hoerling, M., Ji, M., Leetmaa, A., and Sardeshmukh, P., 1996, Assessing GCM's suitability for making seasonal predictions: Journal of Climatology, v. 9, p. 115–129, doi: 10.1175/1520-0442(1996)009<0115:AAGSFM>2.0.CO;2.

Kuzmin, Y.V., Burr, G.S., and Jull, A.J.T., 2001, Radiocarbon reservoir correction ages in the Peter the Great Gulf, Sea of Japan, and eastern coast of the Kunashir, Southern Kuriles (Northwestern Pacific): Radiocarbon, v. 43, p. 477–481.

MacCracken, M.C., Budyko, M.I., Hecht, A.D., and Izrael, Y.A., eds., 1990, Prospects for future climate, Prepared under the auspices of the US/USSR Agreement on Protection of the Environment: Chelsea, Michigan, Lewis Publishers, XIII + 270 p.

Manabe, S., and Stouffer, R.J., 1988, Two stable equilibria of a coupled ocean-atmosphere model: Journal of Climatology, v. 1, p. 841–866, doi: 10.1175/1520-0442(1988)001<0841:TSEOAC>2.0.CO;2.

Marchitto, T.A., Jones, G.A., Goodfriend, G.A., and Weidman, C.R., 2000, Precise temporal correlation of Holocene mollusk shells using sclerochronology: Quaternary Research, v. 53, p. 236–246, doi: 10.1006/qres.1999.2107.

Marshall, J.F., and McCulloch, M.T., 2002, An assessment of the Sr/Ca ratio in shallow water hermatypic corals as a proxy for sea surface temperature: Geochimica et Cosmochimica Acta, v. 66, p. 3263–3280, doi: 10.1016/S0016-7037(02)00926-2.

Matsushima, Y., 1984, Shallow marine molluscan assemblages of postglacial period in the Japanese Islands—Its historical and geographical changes induced by the environmental changes: Bulletin of the Kanagawa Prefectural Museum, v. 15, p. 37–109.

Morton, B., 1970, The tidal rhythm and rhythm of feeding and digestion in *Cardium edule:* Journal of the Marine Biology Association of the United Kingdom, v. 50, p. 488–512.

Mutvei, H., Westermark, T., Dunca, E., Carell, B., Forberg, S., and Bignert, A., 1994, Methods for the study of environmental changes using the structural and chemical information in molluscan shells: Bulletin de l'Institut océanographique, Monaco, no. spécial 13, p. 163–186.

Pannella, G., 1976, Tidal growth patterns in recent and fossil mollusk bivalve shells: A tool for the reconstruction of paleotides: Die Naturwissenschaften, v. 63, p. 539–543, doi: 10.1007/BF00622786.

Pannella, G., and MacClintock, C., 1968, Biological and environmental rhythms reflected in molluscan shell growth: Paleontological Society, Memoir, v. 42, p. 64–81.

Pflaumann, U., and Jian, Z., 1999, Modern distribution patterns of planktonic foraminifera in the South China Sea and western Pacific: A new transfer technique to estimate regional sea-surface temperatures: Marine Geology, v. 156, p. 41–83, doi: 10.1016/S0025-3227(98)00173-X.

Rao, K.P., 1954, Tidal rhythmicity of rate of water propulsion in *Mytilus,* and its modifiability by transplantation: The Biological Bulletin, v. 106, p. 353–359.

Richardson, C.A., Crisp, D.J., and Runham, N.W., 1980, Factors influencing shell growth in *Cerastoderma edule:* Philosophical Transactions of the Royal Society of London, Ser. B, v. 210, p. 513–531.

Sakaguchi, Y., 1989, Some pollen records from Hokkaido and Sakhalin: Bulletin of the Department of Geography, University of Tokyo, v. 15, p. 1–31.

Sato, S., 1997, Shell microgrowth patterns of bivalves reflecting seasonal change of phytoplankton abundance: Paleontological Research, v. 1, p. 260–266.

Sato, S., 1999, Temporal change of life-history traits in fossil bivalves: An example of *Phacosoma japonicum* from the Pleistocene of Japan: Palaeogeography, Palaeoclimatology, Palaeoecology, v. 154, p. 313–323, doi: 10.1016/S0031-0182(99)00106-6.

Schöne, B.R., 2003, A 'clam-ring' master-chronology constructed from a short-lived bivalve mollusc from the northern Gulf of California, USA: The Holocene, v. 13, p. 39–49, doi: 10.1191/0959683603hl593rp.

Schöne, B.R., Goodwin, D.H., Flessa, K.W., Dettman, D.L., and Roopnarine, P.D., 2002a, Sclerochronology and growth of the bivalve mollusks *Chione (Chionista) fluctifraga* and *C. (Chionista) cortezi* in the northern Gulf of California, Mexico: The Veliger, v. 45, p. 45–54.

Schöne, B.R., Lega, J., Flessa, K.W., Goodwin, D.H., and Dettman, D.L., 2002b, Reconstructing daily temperatures from growth rates of the intertidal bivalve mollusk *Chione cortezi* (northern Gulf of California, Mexico): Palaeogeography, Palaeoclimatology, Palaeoecology, v. 184, p. 131–146, doi: 10.1016/S0031-0182(02)00252-3.

Schöne, B.R., Tanabe, K., Dettman, D.L., and Sato, S., 2003, Environmental controls on shell growth rates and $\delta^{18}O$ of the shallow-marine bivalve mollusk *Phacosoma japonicum* in Japan: Marine Biology, v. 142, p. 473–485.

Schöne, B.R., Oschmann, W., Tanabe, K., Dettman, D.L., Fiebig, J., Houk, S.D., and Kanie, Y., 2004, Holocene seasonal environmental trends at Tokyo Bay, Japan, reconstructed from bivalve mollusk shells—Implications for changes in the East Asian monsoon and latitudinal shifts of the Polar Front: Quaternary Science Reviews, v. 23, p. 1137–1150, doi: 10.1016/j.quascirev.2003.10.013.

Southon, J., Kashgarian, M., Fontugne, M., Metivier, B., and Yim, W.W.-S., 2002, Marine reservoir corrections for the Indian Ocean and Southeast Asia: Radiocarbon, v. 44, p. 167–180.

Stuiver, M., and Reimer, P.J., 1993, Extended 14C database and revised CALIB radiocarbon calibration program: Radiocarbon, v. 35, p. 215–230.

Stuiver, M., Reimer, P.J., Bard, E., Beck, W.E., Burr, G.S., Hughen, K.A., Kromer, B., McCormac, F.G., v.d. Plicht, J., and Spurk, M., 1998a, INTCAL98 radiocarbon age calibration 0–24,000 BP: Radiocarbon, v. 40, p. 1041–1083.

Stuiver, M., Reimer, P.J., and Braziunas, T.F., 1998b, High-precision radiocarbon age calibration for terrestrial and marine samples: Radiocarbon, v. 40, p. 1127–1151.

Stuiver, M., Reimer, P.J., and Reimer, R.W., 2000, CALIB 4.3 [WWW program and documentation], http://www.calib.org.

Takahara, H., Sugita, S., Harrison, S.P., Miyoshi, N., Morita, Y., and Uchiyama, T., 2000, Pollen-based reconstructions of Japanese biomes at 0, 6000 and 18,000 yr BP: Journal of Biogeography, v. 27, p. 665–683, doi: 10.1046/j.1365-2699.2000.00432.x.

Tanabe, K., 1988, Age and growth rate determinations of an intertidal bivalve, *P. japonicum,* using internal shell increments: Lethaia, v. 21, p. 231–241.

Ternois, Y., Kawamura, K., Ohkouchi, N., and Keigwin, L., 2000, Alkenone sea surface temperature in the Okhotsk Sea for the last 15 kyr: Geochemical Journal, v. 34, p. 283–293.

Tsukada, M., 1986, Altitudinal and latitudinal migration of *Cryptomeria japonica* for the past 20,000 years in Japan: Quaternary Research, v. 26, p. 135–152, doi: 10.1016/0033-5894(86)90088-8.

Turekian, K.K., Cochran, J.K., Nozaki, Y., Thompson, I., and Jones, D.S., 1982, Determination of shell deposition rates of *Arctica islandica* from the New York Bight using natural ^{228}Ra and ^{228}Th and bomb-produced ^{14}C: Limnology and Oceanography, v. 27, p. 737–741.

Watanabe, T., Winter, A., and Oba, T., 2001, Seasonal changes in sea surface temperature and salinity during the Little Ice Age in the Caribbean Sea deduced from Mg/Ca and $^{18}O/^{16}O$ ratios in corals: Marine Geology, v. 173, p. 21–35, doi: 10.1016/S0025-3227(00)00166-3.

Wefer, G., and Berger, W.H., 1991, Isotope paleontology: Growth and composition of extant calcareous species: Marine Geology, v. 100, p. 207–248, doi: 10.1016/0025-3227(91)90234-U.

Williams, D.F., Arthur, M.A., Jones, D.S., and Healy-Williams, N., 1982, Seasonality and mean annual sea surface temperatures from isotopic and sclerochronological records: Nature, v. 296, p. 432–434, doi: 10.1038/296432a0.

Yasuda, Y., 1983, Climatic variations since the Last Glacial Age as seen from various sedimentological analyses: Climatological Notes, v. 147, p. 47–60.

Manuscript Accepted by the Society 19 April 2005

Printed in the USA

Geological Society of America
Special Paper 395
2005

Orbital forcing of tropical water balance inferred from sulfur speciation in lake sediments

Germán Mora
Department of Geological and Atmospheric Sciences, Iowa State University, Ames, Iowa 50011, USA

Linda Hinnov
Morton K. Blaustein Department of Earth and Planetary Sciences, Johns Hopkins University, Baltimore, Maryland 21218, USA

ABSTRACT

Although water balance in terrestrial settings is an important climate parameter, relatively few proxies are available for reconstructing effective moisture. Here, we investigate the possibility of using sulfur speciation as a proxy for paleorainfall in sediments recovered from a large paleolake in the Bogota Basin, Colombia and spanning the past 660 k.y. Samples were digested with a mild acidic solution to extract acid-soluble sulfur minerals. Sulfur species extracted through this procedure included monosulfides, whereas sulfur in the remaining acid-insoluble fraction was found to be bound to organosulfur compounds. Monosulfide/total sulfur ratios range from 0.2 to 0.9 in the studied sediments and exhibit a cyclic distribution with depth. Low (<0.4) monosulfide/total sulfur ratios are characteristic of glacial intervals, whereas interglacial intervals exhibit both low and high ratios. Because partitioning of sulfur between monosulfides and organosulfur compounds depends on iron availability, we interpret that elevated iron delivery to the paleolake occurs at relatively high precipitation rates and results in the sequestration of sulfur in monosulfide minerals. Conversely, incorporation of sulfur into organic matter occurs at low precipitation rates when iron supply is low. Time-series analysis of sulfur ratios reveals the influence of orbital parameters (i.e., eccentricity, obliquity, and precession). Given that rainfall in the tropics is primarily associated with the passage of the Intertropical Convergence Zone, we conclude that orbital forcing exerts a significant control on the intensity or zonality of the Intertropical Convergence Zone that ultimately affects rainfall patterns in the Colombian Andes.

Keywords: tropics, Pleistocene, Milankovitch, rainfall.

INTRODUCTION

The response of tropical systems to glacial/interglacial fluctuations during the Quaternary is still largely unknown. Although marine and terrestrial records in the tropics show rhythmic fluctuations with periodicities centered at the main orbital parameters, the timing and amplitude of these fluctuations varies from site to site (e.g., Hays et al., 1976; Mix et al., 1986). Consequently, studies of these dissimilar responses in time and intensity could provide valuable information on the mechanisms of global change, allowing an evaluation of the role of the tropics in triggering and/or responding to climate change. This role is particularly important considering that large amounts of energy and moisture are exported from low- to mid-latitudes. An indicator of the dynamics of the transfer of moisture and energy is tropical rainfall patterns. For example, a dry spell in the tropics is likely to result in less latent heat transferred to high latitudes, thereby promoting a cooling event. One avenue to understanding the role of

Mora, G., and Hinnov, L., 2005, Orbital forcing of tropical water balance inferred from sulfur speciation in lake sediments, *in* Mora, G., and Surge, D., eds., Isotopic and elemental tracers of Cenozoic climate change: Geological Society of America Special Paper 395, p. 33–42, doi: 10.1130/2005.2395(04). For permission to copy, contact editing@geosociety.org.

tropical rainfall patterns is through assessments of past changes in precipitation patterns and evaluations of their effect on global climate change.

Despite their importance, reconstructions of past rainfall patterns in tropical regions are scant and provide contradictory interpretations. For example, while some pollen records suggest drier conditions in Amazonia for the last glacial interval (e.g., van der Hammen and Absy, 1994), other pollen records indicate climate conditions similar to those prevailing during the current interglacial period (e.g., Haberle and Maslin, 1999; Colinvaux et al., 2000). Contrary to records in Amazonia that provide apparent contradictory information of water balance, records in the Caribbean consistently show drier conditions during the Last Glacial Maximum (LGM) (Hodell et al., 1991; Peterson et al., 1991). Because the position of the Intertropical Convergence Zone controls precipitation rates in the tropics, the dry spell recorded in the Caribbean for the LGM is explained in terms of a southern shift in the mean annual position of the Intertropical Convergence Zone. This hypothesis is based on modern observations indicating that increased warming in South America during boreal winter produces an intensification of the northeastern trade winds and a displacement of the southeastern trade winds into the Southern Hemisphere. The combined effect displaces the Intertropical Convergence Zone into the Southern Hemisphere, creating a dry season over Africa and the northern neotropics. In South America, the southward displacement of the southeastern trade winds is accentuated during boreal winter, driving the Intertropical Convergence Zone to ~20°S. Nonetheless, the southeastern trade winds remain in the Northern Hemisphere for much of the year as a result of the irregular configuration of the continents (Riehl, 1979).

The forcing mechanism suggested for the proposed shifts in the annual position of the Intertropical Convergence Zone remains controversial. Whereas Martin et al. (1997) propose a local mechanism in which changes in low-latitude insolation are the main factor controlling the position and intensity of the Intertropical Convergence Zone over South America, Harris and Mix (1999) postulate that changes in the location of the Intertropical Convergence Zone are triggered by an oceanic circulation mechanism forced by insolation in boreal temperate regions. Alternatively, Baker et al. (2001) propose a combined effect of local, low-latitude insolation and sea-surface temperature gradients along the equatorial Atlantic in controlling moisture balance over tropical South America.

Our goal is to document changes in water balance in the tropical Andes over glacial/interglacial periods. This documentation can provide valuable information to test whether local insolation or extratropical forcing controls the position of the Intertropical Convergence Zone. The best archives of water balance changes come from lacustrine sequences, which provide high-resolution paleoclimate records, if accumulation rates are sufficiently high to allow for assessing processes at the rate of glacial/interglacial oscillations. Unfortunately most of the available continental records from South America are short and discontinuous, causing Quaternary climate studies of the neotropics to be based on marine records (e.g., de Menocal et al., 1993; Harris and Mix, 1999). The exception, however, is the nearly continuous Pleistocene lacustrine record of the Bogota basin (Colombia) in northern South America (Fig. 1). This lacustrine sequence developed in an intermontane depression of the tropical Andes (lat. 4°30′–5°15′N) at an altitude of ~2550 m, near the current Andean tree line. Two cores (Funza-I and Funza-II) were retrieved from the center of the basin. Pollen data from these cores provide evidence for climate-induced shifts in the elevation of the upper tree line of the Andes during glacial/interglacial intervals (Hooghiemstra and Ran, 1994). Frequency analysis of the pollen record suggests a linear linkage between high-latitude orbital forcing and neotropical alpine climate, but the cause-and-effect mechanism remains speculative. In this study we use the Funza record to reconstruct effective moisture conditions for northern South America during the past 660 k.y., relying on time-series analysis in an attempt to unravel the mechanisms responsible for variations in effective moisture level and to test the influence of local versus extratropical forcing on the level of repeated climate fluctuations in the Bogota basin during the Pleistocene.

To reconstruct effective moisture levels in the Bogota Basin using the Funza core, we employ the relative abundance of sulfur species in the sediments as a proxy for rainfall conditions because the partitioning of sulfur species in lake sediments depends on

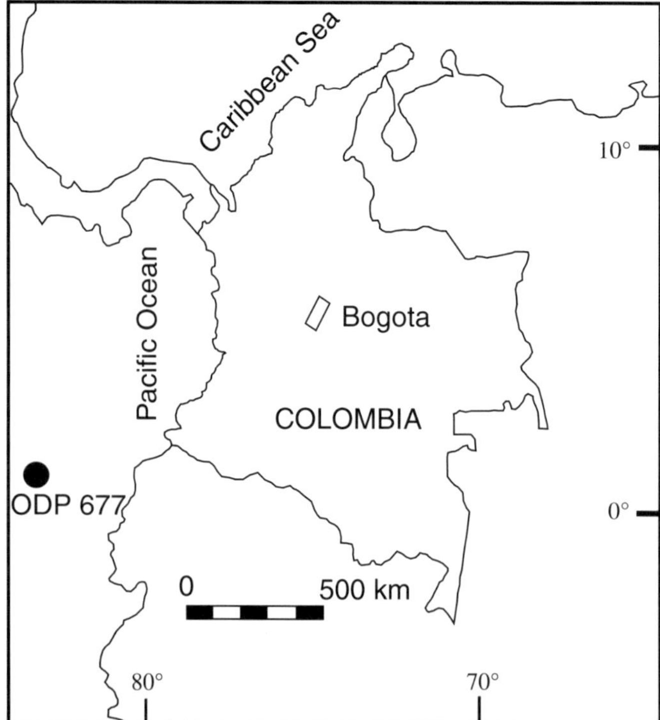

Figure 1. Map of northern South America, showing the location of the Funza cores (in the Bogota Basin) and the location of Ocean Drilling Program Site 677.

iron availability on a first-order basis. H_2S produced as a result of the activity of sulfur-reducing bacteria could react with iron to form monosulfides or could be oxidized to either elemental sulfur or polysulfides and react with organic matter to create new sulfur-bearing organic compounds. Kinetics, however, favors the formation of mineral sulfides over organosulfur compounds (Pyzik and Sommer, 1981). Because rainfall intensity controls iron delivery from watersheds to lakes, a limited supply of iron would promote the formation of organosulfur compounds. This relationship between rainfall-influenced iron delivery and diagenetic formation of sulfur compounds is the basis for a potentially useful paleoclimatic proxy. The rationale behind the use of sulfur species abundances in lake sediments to reconstruct rainfall patterns is explained in the following section.

SULFUR IN FRESH-WATER LAKES

Sources of sulfur in lakes include dissolved sulfate and sulfur bound to organic compounds. In pristine lakes, almost all dissolved sulfate comes from the watershed as a result of weathering of sulfur-bearing rocks and the oxidation of organic sulfur in soils (Cook and Kelly, 1992). Sulfur bound to organic compounds is primarily linked to aquatic plants mainly in the form of proteins but other important compounds include sulfolipids, sulfated polyssacharides, sulfated esters, and vitamins (e.g., biotin and thiamine). Sulfur occurs in four major oxidation states in organic compounds, including organic sulfides, di- and polysulfides, sulfate esters, sulfonates, and sulfoxides. Sulfur is also present in thiophenes and heterocyclic structures (Ferdelman et al., 1991; Vairavamurthy et al., 1994; Xia et al., 1998). The assimilation of sulfur by aquatic plants mainly occurs through the incorporation of sulfate and the subsequent assimilatory reduction and incorporation of carbon-bonded sulfur into amino acids.

In addition to aquatic plants, bacterial organisms synthesize sulfur-bearing amino acids via sulfate reduction. This process mainly occurs in sediments where dissolved sulfate diffuses into the sediments as a result of a concentration gradient that exists between pore waters and lake waters. Once dissolved sulfate enters lake sediments, sulfate-reducing bacteria (SRB) utilize the available sulfate to oxidize organic compounds. Sulfate reduction occurs under anoxic conditions in lake sediments as indicated by field data and culture experiments, showing that the presence of oxygen inhibits sulfate reduction by bacteria (Canfield and Des Marais, 1991; Sass et al., 1997). In sediments, sulfate also originates during the hydrolysis of sulfate-esters, but this is a minor source of sulfate in lake sediments, accounting for less than 10% of the total sulfate pool available to SRB (King and Klug, 1982). Sulfate concentrations in fresh-water lakes are typically low, ranging from 300 μM to as high as 800 μM in meso- and eutrophic lakes. These low concentrations limit the diffusion of sulfate in the sediments, causing the zone of sulfate reduction to be commonly restricted to the upper 10 cm of the sediment column (Molongoski and Klug, 1980; Smith and Klug, 1981). Despite low concentrations and limited penetration, measured sulfate reduction rates in fresh-water lakes can achieve values similar to those reported in marine sediments (Lovley and Klug, 1986; Sinke et al., 1992). These high rates result from intense sulfur recycling in lake pore waters as reduced sulfur produced by SRB activity can migrate to oxic layers and be oxidized back to sulfate (Jorgensen, 1990; Roden and Tuttle, 1993).

Ultimately the byproducts of SRB-mediated sulfate can either be sequestered in solid inorganic phases or in organic compounds (Cook and Kelly, 1992; Urban and Monte, 2001). As a result, sedimentary sulfur in fresh-water lakes occurs in two main phases: organosulfur compounds and sulfide minerals. A number of studies have addressed the incorporation of sulfur into organic compounds (for a review, see Sinninghe Damsté and de Leeuw, 1990). Laboratory experiments indicate that reduced sulfur is included into functionalized lipids and possible carbohydrates through intra- and intermolecular reactions (e.g., Schouten et al., 1994; Sinninghe Damsté et al., 1998). Field studies (e.g., Wakeham et al., 1995; Putschew et al., 1996; Werne et al., 2000) indicate that the incorporation of reduced sulfur into organic matter requires (1) a suitable organic substrate prone to react with reduced sulfur, (2) anoxic conditions to ensure that sulfur remains reduced, and (3) a limited supply of iron to ensure that sufficient sulfide is available for reaction (Canfield, 1989) because iron sulfide precipitation appears to occur more readily than the incorporation of sulfur into organic matter (Hartgers et al., 1997).

Organosulfur compounds are typically the most abundant pool of sedimentary sulfur in lake sediments, particularly in eutrophic lakes (King and Klug, 1982; Landers and Mitchell, 1988; Kleeberg, 1997). Marnette et al. (1992) demonstrated that higher abundances of organosulfur compounds relative to those of sulfides occur when pools of sedimentary iron are low and pore waters are undersaturated with respect to iron sulfides. This finding suggests that the partitioning of sulfur between organic and inorganic phases in lake sediments is primarily dependent on iron availability. Consequently the abundance of iron available to reduced sulfur species determines the fate of sulfur in lake sediments (e.g., Cook, 1984; Cook and Kelly, 1992; Hartgers et al., 1997; Friese et al., 1998).

Iron hydroxides are the most important source of iron in lakes, coming from the watershed (Wetzel, 1983). In some lakes, extremely low amounts of iron have been reported (e.g., Wetzel, 1972). Under these conditions, monosulfide formation is restricted by limited iron availability in lake pore waters. The abundance of iron hydroxides in aquatic systems of tropical regions is closely linked to precipitation rates (Wetzel, 1972). Reduced rainfall limits weathering rates, thereby reducing the delivery of iron hydroxides to lake bodies. As a result, organosulfur compounds are more abundant in lake sediments relative to monosulfide minerals. In contrast, enhanced rainfall intensifies weathering rates, increases iron hydroxide delivery to lakes, and results in the preferential formation of iron monosulfides over organosulfur compounds.

The close linkage between rainfall intensity and iron delivery and abundance in lake systems produces a proxy that can be used to assess past precipitation patterns. This relationship is used in this study in an effort to assess past water balance levels in the paleolake that occupied the Bogota Basin in the Colombian Andes. Consequently relative abundances of sulfur species (metal sulfides and organosulfur compounds) were determined in the Funza-II core. It is expected that drier periods should diminish iron delivery, limit the sulfide formation in lake sediments, and result in enhanced preservation of organosulfur compounds. In contrast, wetter periods should enhance weathering rates and the consequent delivery of iron oxides to the paleolake of the Bogota basin, promoting sulfide precipitation in lake pore water and limiting the formation of organosulfur compounds (Mora et al., 2002).

METHODS

Samples collected for this study were taken from the Funza-II core. Upon retrieval in 1988 (Hooghiemstra and Ran, 1994), a split of the core was stored under cool (4 °C), dark, and low-humidity conditions, and a split was stored under freezing conditions. Sediment samples were collected at 1 m intervals from both splits to assess the effect of long-term storage on the abundance of sulfur and carbon in the samples. Collected samples were brought to the laboratory, dried, and powdered. About 0.25 g of dried, powdered sediment was acidified with dilute HCl at room temperature to remove carbonates. The insoluble residue was neutralized, dried, and then combusted in a LECO C/S 244 analyzer to determine organic carbon (C_{org}) and acid-insoluble sulfur (S_{ai}) contents. Total sulfur (S_{tot}) and total carbon (C_{tot}) were determined by directly combusting dried, unacidified, powdered samples. The difference between S_{tot} and S_{ai} corresponds to the acid-soluble sulfur (S_{as}) fraction. The results of the geochemical analyses for the samples from a split kept under freezing conditions differ by less than 5% with respect to values obtained from samples stored at cool conditions. Analytical reproducibility ranged from ±0.1% to ±2.8% for C_{org} values and from ±0.7% to ±7.1% for S_{tot} values.

S_{ai} encompasses both sulfur bound to both organic compounds and pyrite. To determine the pyrite content in the samples, chromium reduction analyses (Canfield et al.,1986; Brüchert and Pratt, 1996) were performed on samples exhibiting high S_{ai} concentrations. Briefly, aliquots of sediments were reacted with 6N HCl under N_2 gas to evolve acid-volatile monosulfides. The residue was reacted with a 1:1 mixture of 12N HCl and 1M $CrCl_2$ under N_2 gas. Pyrite present in the residue transforms into H_2S, which is carried by a stream of N_2 through a buffer solution of citric acid and NaOH (pH = 3.5) and into a 0.1M $AgNO_3$ solution. Evolved H_2S precipitates as Ag_2S, which is subsequently filtered, rinsed, and dried.

The chronology of the core was established on the basis of radiocarbon dates in the upper part of the core (Mommersteeg, 1998), and fission-track dates determined dates in three ash layers (Andriessen et al., 1993). Spectral analysis of the pollen record obtained for the Funza core revealed that variations in pollen abundance were coincident with orbital parameters (Hooghiemstra et al., 1993), allowing for tuning of the Funza record with well-dated marine sequences (Mommersteeg, 1998). The chronology of the Funza record was further refined using the sulfur data. For that, the 122.6-m-long, one-meter sampled series of sulfur variations was analyzed using multiple taper spectral estimation (Thomson, 1982) with the aid of the *Analyseries* software (Paillard et al., 1996). In addition, amplitude spectrograms (Matlab™) were applied to investigate time-frequency changes related to unstable sedimentation rates and signal frequencies.

RESULTS

The Funza-II core is composed mainly of massive, sometimes faintly mottled, clays that lack distinct burrow structures or fine laminations. Thin beds of fine- to medium-grained sand and occasional thin layers of volcanic ash are present. Consistent with previous lithological descriptions of the core (Hooghiemstra and Ran, 1994), we did not observe the development of pedogenic or leached intervals, suggesting a continuous accumulation of sediments. Values of C_{org} for the study interval of the Funza-II core were within the analytical uncertainty of those for C_{tot}, suggesting the absence of carbonates in the sediments.

Values of C_{org} in the 5–120 m interval of the Funza-II core show an irregular pattern (Fig. 2) with concentrations ranging from 0.22 wt% to 14.6 wt% and an average value of 3.3 wt%. In the upper part (5–53 m) of the core, C_{org} values are typically 2.2 wt% higher than those measured in the lower part of the core.

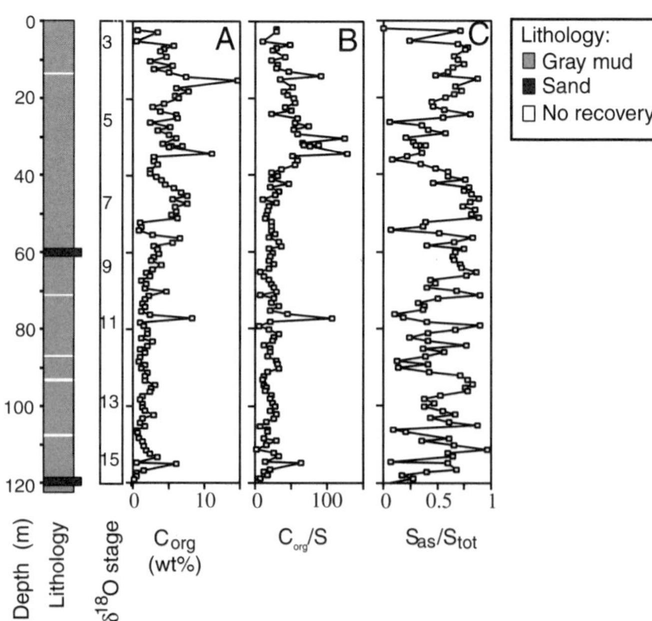

Figure 2. (A) Values of C_{org} in the 5 to 120 m interval of the Funza-II core. (B) Ratio of organic carbon to total sulfur (C_{org}:S). (C) Ratios of S_{as}:S_{tot}, the precipitation proxy.

Total sulfur concentrations in the Funza record fall within those typically measured in fresh-water lakes (Berner et al., 1979), ranging from 0.02 wt% to 0.803 wt%. These values are significantly lower than those typically found in modern marine sediments, indicating a fresh-water system (Fig. 3). Ratios of organic carbon to total sulfur (C_{org}:S) range from 5 to 127, with typical values in the range of 15–30. Ratios of S_{as}:S_{tot}, where S_{as} = S_{tot} – S_{ai}, in Funza sediments are on average 0.5 with a maximum of ~0.95. Chromium reduction analyses reveal the absence of pyrite in the studied sediments.

DISCUSSION

Ratios of C_{org}:S_{tot} in the Funza record are typically 50 units lower than those typically found in plankton (~100, Baker et al., 1989), suggesting an efficient sulfur retention or carbon loss in sediments given that the most important source of sedimentary sulfur in fresh-water lakes is organic matter. Sulfur is retained in sediments through two major pathways: sulfide precipitation or formation of organosulfur compounds (Nriagu and Soon, 1985; Mitchell et al., 1990). Unfortunately current analytical techniques are inaccurate for determining the concentration of sulfur substrates of these two sedimentary sulfur pools. To circumvent this problem, extraction techniques are performed on sediments to distinguish broadly between different sedimentary sulfur forms. Acid treatment dissolves some sulfur components, yielding the operationally defined acid-soluble sulfur (S_{as}). It has been found that S_{as} generally corresponds to metallic monosulfides (e.g., FeS), with a metal/sulfur relationship of ~1:1 (Jorgensen, 1977). In this study, samples were treated under mildly acidic conditions (1 N HCl) at room temperature (see Methods section above). Rice et al. (1993) suggest that this treatment may be insufficient to fully dissolve some monosulfides. Therefore, measured S_{as} in the Funza core should be considered as a minimum value. After acid treatment, the insoluble fraction (referred to as acid-insoluble sulfur, S_{ai}) corresponds to a combination of pyrite (FeS_2), elemental sulfur, polysulfides, and organic sulfur (Canfield et al., 1986). Polysulfide concentration in fresh-water lacustrine sediments is, however, typically low or below detection limits (Nriagu and Soon, 1985). Chromium reduction analysis on Funza sediments reveals the absence of pyrite. Consequently, Funza S_{ai} values correspond to the abundance of organic sulfur and possibly elemental sulfur. The moderate correlation observed between S_{ai} and C_{org} in Funza-II (Fig. 3) supports the hypothesis that organosulfur compounds are the main constituent of S_{ai}.

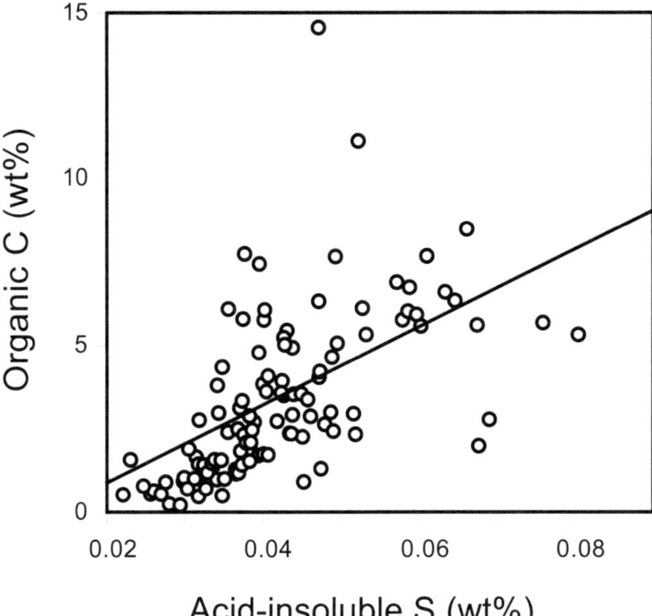

Figure 3. Correlation between S_{ai} and C_{org} in Funza-II.

Although pyrite is a common mineral phase in lake sediments, experimental results (e.g., Benning et al., 2000) and observations in modern lake sediments (e.g., Bates et al., 1995) indicate that pyrite could be absent in sediments when anoxic conditions persist in the sedimentary column and sulfur is not abundant. We hypothesize that these conditions occurred in the Funza sediments because the paleolake was likely depleted in sulfur as indicated by low sedimentary sulfur abundances and was likely anoxic because organic carbon concentrations are relatively high. It has generally been accepted that an oxidant is required to produce pyrite from iron monosulfides (Berner, 1970; Luther, 1991; Benning et al., 2000). Consequently, the preservation of monosulfides in sediments has been attributed to the spatial and temporal decoupling between potential oxidants and precipitated iron monosulfides (e.g., Middelburg, 1991; Lyons, 1997). We hypothesize, therefore, that high sedimentation rates in the Funza record would likely have prevented the possible advection of potential oxidants from the water column, thereby preserving iron monosulfides in the sediments.

Ratios of S_{as}:S_{tot}, where S_{as} = S_{tot} – S_{ai}, in Funza sediments show a rhythmic distribution with depth (Figs. 2 and 4). Our results indicate that S_{as}:S_{tot} ratios for the Funza-II core are higher than average for lacustrine sediments (Berner et al., 1979). These measured ratios could be even higher if the monosulfides were not fully dissolved as discussed above. These high S_{as}:S_{tot} ratios suggest that most of the sulfur is in the form of monosulfides precipitated under reducing conditions as a result of bacterial activity. SRBs utilize dissolved sulfate as an electron acceptor to oxidize organic matter, transforming sulfate into HS^- or H_2S. Reduced sulfur species can subsequently react with ferrous cations to form monosulfides (Berner et al., 1979).

In addition to high S_{as}:S_{tot} ratios, some intervals in the Funza core also exhibit low (<0.5) ratios, indicating a more significant contribution of S_{ai} to the sedimentary sulfur. The absence of pyrite, the likely absence of polysulfides, and the moderate correlation between S_{ai} and C_{org} in Funza-II (Fig. 3) lead us to suggest that organosulfur compounds are the main constituents of S_{ai} in the studied sediments.

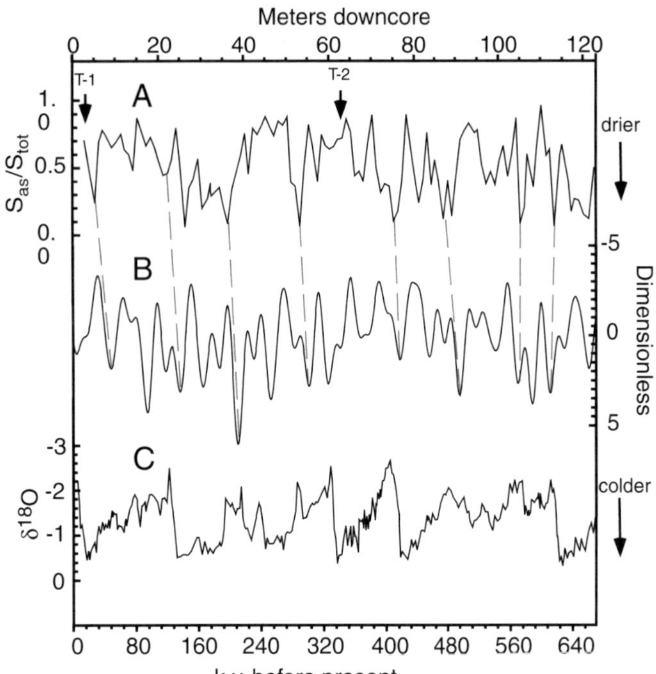

Figure 4. Comparison of the Funza-II precipitation proxy, Earth's orbital parameters, and the paleoclimate series from Ocean Drilling Program Site 677. (A) The S_{as}:S_{tot} 1 m sampled series with T-1 the radiocarbon date of 25 ± 3 ka obtained at 2.8 m, and T-2 the fission-track date of 250 ± 100 ka obtained from an ash layer at 67 m (see text). Depth axis in meters is at the top. (B) Standardized eccentricity, tilt, precession (ETP) curve, after Laskar et al. (1993), referred to the time axis at the bottom. (C) Planktonic marine isotope record of Ocean Drilling Program 677, tuned to an orbitally forced glaciation model (Shackleton et al. 1990), also referred to the time axis at the bottom.

Organosulfur compounds encompass both compounds originally present in settling organic matter and those formed in the sediments during diagenesis. The diagenetic incorporation of sulfur in organic matter occurs as inorganic sulfur species, formed under mildly reducing conditions by SRBs, react with sedimentary organic matter and form new organosulfur compounds (Vairavamurthy and Mopper, 1987; LaLonde, 1990). Consequently we interpret that low S_{as}:S_{tot} ratios result under limited iron inputs, whereas enhanced iron delivery to the lake increases the retention of sulfur as monosulfides, resulting in high S_{as}:S_{tot} ratios.

Low amounts of organic matter delivered to the sediments or low concentrations of dissolved sulfate could also explain the observed elevated abundance of organosulfur compounds in the Funza record relative to the total sedimentary sulfur pool. Reduced availability of organic matter and/or sulfate could suppress sulfate-reduction rates, resulting in a reduction of H_2S production and the consequent decrease of monosulfide precipitation. Consequently lower sulfate reducing rates lead to low S_{as}:S_{tot} ratios due to the limited formation of monosulfide in the sediments. Decreased rates of sulfate reduction are typically associated with high (110–160) C_{org}:S_{tot} ratios as a result of little bacterial activity and poor retention of sulfides (David and Mitchell, 1985; Cook et al., 1990). The correlation between elevated C_{org}:S_{tot} ratios and low S_{as}:S_{tot} ratios was weak in the Funza record. In fact, our geochemical results indicate the opposite. Intervals with presumably enhanced organosulfur compounds (i.e., low S_{as}:S_{tot} ratios) typically exhibit low C_{org}:S_{tot} ratios (Fig. 2), making the possibility of decreased sulfate-reduction rates unlikely. A few intervals in the Funza record do show high (~100) C_{org}:S_{tot} ratios linked to low S_{as}:S_{tot} ratios (Fig. 2). Geochemical and petrographic characterization of these intervals indicates the significant presence of terrestrially derived organic matter (Mora et al., 2002). We hypothesize that the combined effect of abundant terrestrial organic matter and the refractory nature of this organic detritus limited sulfate reduction rates in lake sediments, resulting in high sedimentary C_{org}:S_{tot} ratios and low S_{as}:S_{tot}. With these few intervals as exceptions, the sulfur record in the Funza core reflects changes in iron delivery.

Given that rainfall strongly influences lacustrine iron inputs, we then propose the use of S_{as}:S_{tot} ratios as a proxy for precipitation in the Bogota basin. Consequently we interpret that lower precipitation regimes in the basin resulted in relatively high S_{as}:S_{tot} ratios due to the limited input of iron from the watershed and the consequent lower monosulfide formation with respect to total sulfur. Conversely, increased chemical weathering caused by high precipitation rates would have enhanced the delivery of iron to the lake. Iron would preferentially have reacted with bacteria-produced H_2S in the lake sediments under reducing conditions, leading to monosulfide precipitation over formation of new organosulfur compounds.

Time-Series Analysis

A basic time scale for the Funza-II core was established previously by a radiocarbon date near the top of the core (Mommersteeg, 1998) and fission-track dates (250 ± 100 ka at 67 m, and 1.02 ± 0.23 Ma at 303 m; Andriessen et al., 1993). These dates give an average sedimentation rate of 0.257 m/k.y. for the interval 2.8 m to 67 m and 0.306 m/k.y. for the interval 67 m to 122.6 m, for a total time interval of 407 k.y., spanning 25 ka to 432 ka. This chronology places the meter-scale Funza-II S_{as}:S_{tot} variations at time scales of orbitally forced climates. Consequently the Funza-II series was compared to the Earth's orbital parameters and to the well-known surface ocean oxygen isotope record of Ocean Drilling Program Site 677 (see location in Fig. 1). These three series are plotted together in Figure 4, and their corresponding average power and time-frequency spectra are displayed in Figure 5.

The spectrum of the Funza-II series (Fig. 5A) contains power in three well-defined frequency bands at spacings consistent with those of the Earth's orbital parameters (Fig. 5B). The Funza-II spectrogram also reveals that these frequencies shift appreciably through the length of the core, suggesting that variable sedimentation rates affect the record between the dated intervals. In particular, there is a rapid shift of power to very high frequencies near

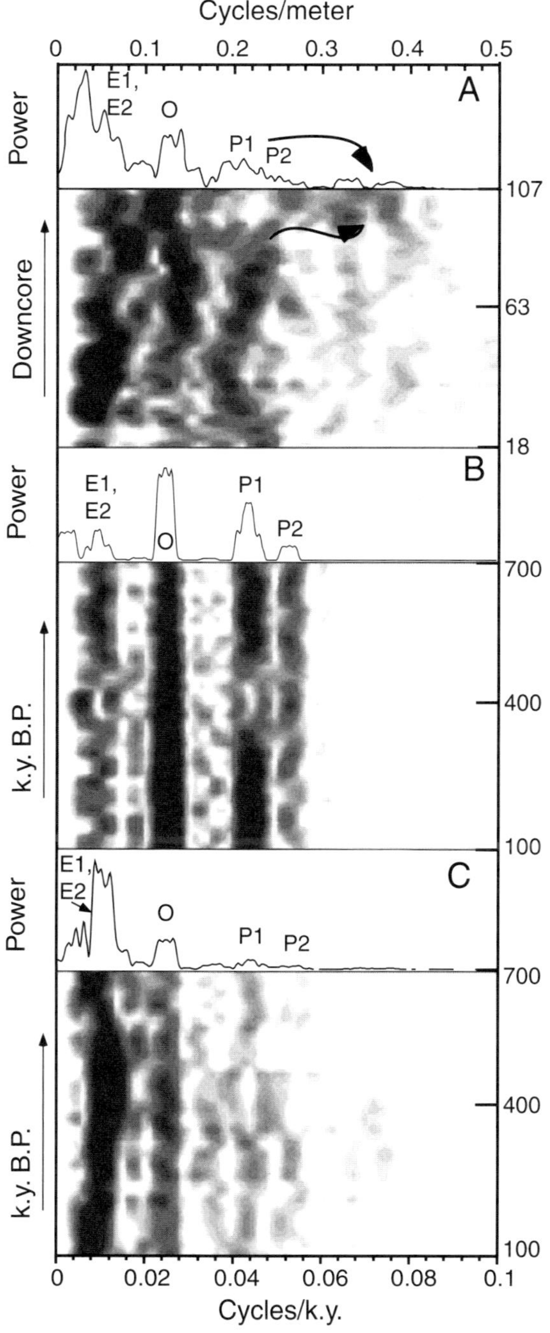

Figure 5. 2π Multitaper power spectra and amplitude spectrograms, where E1, E2 refers to short eccentricity frequency components at 1/(128 k.y.) and 1/(95 k.y.); O is the obliquity frequency component 1/(41 k.y.), P1 is the long precession 1/(23 k.y.), and P2 is the short precession 1/(19 k.y.). (A) The Funza-II precipitation proxy S_{as}:S_{tot} core series, plotted so that the peak at 1/(7.7 m) calibrates to the obliquity frequency component, with the spectrogram calculated with 30 m windowed FFTs over 3 m steps, referred to the frequency axis at the top. (B) ETP curve from 0 to 800 ka, with the spectrogram calculated with 200 k.y. windowed FFTs over 20 k.y. steps, referred to the frequency axis at the bottom. (C) Planktonic marine isotope record of Ocean Drilling Program Site 677 from 0 to 800 ka, with the spectrogram calculated also with 200 k.y. windowed FFTs over 20 k.y. steps and referred to the frequency axis at the bottom.

the very base of the sampled interval (see arrows in Fig. 5A), which we interpret as a sudden decline in sedimentation rates of ~40% that causes variations to become "squeezed." On average, however, the frequencies remain stable enough throughout the sampled core interval to contribute significantly to the three main bands.

The similarities between the Funza-II and ETP (eccentricity, tilt, precession) spectra led us to calibrate the peak at f = 0.13 cycles/meter (the 7.7 m S_{as}:S_{tot} cycle) to the Earth's 41 k.y. obliquity variation (O). This calibration scales the other two peaks to the frequency bands of the short eccentricity (Fig. 5, E1, E2) and the precession index (Fig. 5, P1, P2). This calibration requires an adjustment of the sedimentation rates that had been assumed using the reported radiocarbon and fission-track dates to a somewhat slower rate of 7.7 m/(41 k.y.) = 0.188 m/k.y., and a corresponding expansion of the estimated total time interval for the record from 407 k.y. to 119.8 m/(0.188 m/k.y.) = 637.2 k.y., or from 22 ka to 659 ka. This adjustment falls within the estimated uncertainties of the absolute dates, making it a valid alternative chronology. Plotting the Funza-II series according to this adjusted time scale (Fig. 4A) shows that many of its variations appear to correlate with those of the ETP curve (Fig. 4B). The chief difference is that the Funza-II series has a stronger 100 k.y. scale component than does the ETP curve, possibly related to the short orbital eccentricity (E1, E2; see also Figs. 5A and 5B). This enhanced 100 k.y. frequency is well known in the marine isotope record over the late Quaternary and is also well represented in the Ocean Drilling Program Site 677 isotope record shown in Figures 4C and 5C. The main difference between the Ocean Drilling Program Site 677 and Funza-II records is that the latter has substantially higher power in the precession frequency band, despite the fact that Ocean Drilling Program Site 677 was sampled at a higher time resolution (1–3 k.y. versus the 1 m/(0.188 m/k.y.) = ~5.3 k.y. resolution of Funza-II).

PALEOCLIMATIC SIGNIFICANCE

Our geochemical data reveal significant changes in rainfall patterns in the tropical Andes during glacial/interglacial intervals as suggested by fluctuations in the Funza-II S_{as}:S_{tot} ratios. At present, however, no other continental paleoclimate archive recording the past ~660 k.y. exists in the northern neotropics. Thus to verify that our sulfur proxy for rainfall is consistent with other interpretations, we must rely on shorter records. The sulfur data in the Funza core indicate that the northern tropical Andes experienced drier conditions during the last glaciation (Fig. 4). This is in agreement with unusually high hydrogen and oxygen isotopic composition of pedogenic kaolinite formed in the Bogota basin during the last glaciation, which was interpreted to reflect drier conditions (Mora and Pratt, 2001). Other evidence for arid conditions in the northern neotropics during the last glaciation comes from a ~150 k.y. pollen record in Lake El Valle, Panama (Bush, 2002). During that interval, this record shows an increase in savanna vegetation inferred to result from lower rainfall. Drier conditions in the Caribbean region have also been postulated for the last glaciation on the basis of elevated oxygen isotopic composition of lacustrine ostracods in Haiti (Hodell et al., 1991) and decreased titanium abundances in marine sediments from the Cariaco basin (Haug et al., 2001). Consequently the correspondence between other paleoclimate proxies indicating drier conditions for the northern neotropics during the last glaciation and lower $S_{as}:S_{tot}$ ratios in the Funza core provides additional support for using sulfur data as a proxy for rainfall.

The distribution of $S_{as}:S_{tot}$ ratios with time shows that the lowest values (driest intervals) occur during glacial maxima, whereas relatively high $S_{as}:S_{tot}$ values (wet periods) typically characterize glacial terminations (compare Figs. 4A and 4C). Harris and Mix (1999) similarly concluded that dry periods occurred during ice growth and wet periods during ice melting in the Amazon basin. Their analysis is based on the relative abundance of detrital iron oxides in a one-m.y.-long sequence from the Ceara Rise, Atlantic Ocean. Dry periods inferred from increased hematite abundance were found in intervals associated with ice growth of continental ice sheets in the Northern Hemisphere indicated by elevated oxygen isotopic composition of foraminifera in the same core. In contrast, wet periods (elevated goethite) occurred during ice melt. Another significant finding from this work was that the proportion of oxides in the sediments of the Ceara Rise varied in accordance to orbital parameters, thereby indicating the possible control of orbital forcing on rainfall patterns in the Amazon basin during the Quaternary.

The spectral estimates of $S_{as}:S_{tot}$ ratios in Figure 5 reveal a cyclic character of variations in moisture balance for the Bogota Basin, showing significant peaks with periodicities closely corresponding to those of the orbital parameters. A substantial portion of the total variance is concentrated at the ~41 k.y. period, suggesting a significant influence of the obliquity variation on rainfall patterns in the Bogota Basin. The presence of this cycle occurs also in pollen data from the Funza-I core (Hooghiemstra et al., 1993). Thus, sulfur speciation points toward a feedback mechanism involving high latitude processes because the obliquity parameter dominates insolation at middle and high latitudes and contributes little to insolation at low latitudes (Berger et al., 1992).

Because the Intertropical Convergence Zone controls rainfall patterns in tropical regions, the inferred changes in water balance in the Bogota Basin should have resulted from shifts in the position or intensity of the Intertropical Convergence Zone (Mora and Pratt, 2001). Mayle et al. (2000) invoked latitudinal shifts in the mean annual position of the Intertropical Convergence Zone to explain increased rainfall amount in southern Amazonia during the past 3 k.y. as inferred from increased rain forest elements. Similarly Martin et al. (1997) also propose a shift of the Intertropical Convergence Zone to explain changes in precipitation rates in southeastern Brazil. These authors propose that a northward shift of the Intertropical Convergence Zone during the Last Glacial Maximum caused increased savanna vegetation in southeastern Brazil. A mechanism currently invoked to explain the interpreted shifts in the position of the Intertropical Convergence Zone involves changes in local insolation (Hodell et al., 1991; Seltzer et al., 2000; Baker et al., 2001). According to this hypothesis, insolation at low latitudes, which is fundamentally controlled by precession (Berger et al., 1992), controls the strength of tropical convective activity. In contrast to this proposal, our analysis reveals the presence of a strong obliquity signal in the Funza sulfur data, thereby suggesting the control of high-latitude insolation on the intensity of the Intertropical Convergence Zone. Supporting this suggestion, Harris and Mix (1999) propose that changes in the location of the Intertropical Convergence Zone are triggered by an oceanic mechanism forced by Northern Hemisphere high-latitude insolation. However, uncertainties in the chronology of the Funza core preclude us from convincingly concluding that an extratropical forcing mechanism was responsible for shifts in the Intertropical Convergence Zone during the past 700 k.y.

CONCLUSIONS

We measured sulfur abundances on a unusually long sequence of lake sediments from the neotropics (Funza-II core) to evaluate $S_{as}:S_{tot}$ ratios as a proxy for rainfall variation in northern South America for the past 660 k.y. We propose the use of $S_{as}:S_{tot}$ ratios as a moisture balance proxy in lakes where the available iron is insufficient to sequester dissolved sulfur during the earlier stages of diagenesis near the water-sediment interface. Under this scenario, excess sulfur reacts with organic matter to form significant amounts of organosulfur compounds. Because kinetics favors the formation of iron sulfides over the development of organosulfur compounds (Pyzik and Sommer, 1981), increased supply of iron due to enhanced rainfall results in elevated abundances of iron sulfides with respect to those of organosulfur compounds. It is currently unknown how widespread sedimentary sulfur ratios can be used as a paleoclimatic proxy, but we estimate that a number of fresh-water lakes can meet the criteria outlined here. The Funza record provides a good example because of its relatively low sulfur abundances, the limited availability of iron, and relatively high accumulation rates.

Ratios of $S_{as}:S_{tot}$ in the Funza core reveal a cyclic distribution with depth. We performed a spectral analysis of the sulfur data for the Funza core to assess the frequency of these cycles. The spectrum of the sulfur data reveals a cyclic pattern containing power in three distinct frequency bands. The spacing between these bands is similar to that existing between eccentricity, obliquity, and precession. Consequently we calibrate peaks in the Funza sulfur spectra with the Earth's orbital parameters. If this calibration is correct, it indicates a strong obliquity forcing the distribution of sulfur species in the Funza record. We interpret changes in lacustrine sulfur speciation in terms of changes in iron delivery from the watershed induced by changes in rainfall patterns. Since the Intertropical Convergence Zone controls rainfall patterns in the tropics, we conclude that shifts in the position or intensity of the Intertropical Convergence Zone are possibly related to the Earth's orbital parameters as suggested by the rhythmic distribution of sulfur species in the Funza record. In agreement with our conclusion, Harris and Mix (1999) also propose a possible control of orbital forcing on tropical rainfall patterns. Similar to their findings, the sulfur data reveal the prevalence of dry conditions during glacial maxima and the occurrence of wet periods during glacial terminations.

ACKNOWLEDGMENTS

The authors would like to thank J. Werne and Lisa M. Pratt for reviewing this manuscript, Dr. H. Hooghiemstra and Ingeominas (Colombia) for granting permission to collect samples of the Funza record, and Ana M. Carmo for providing assistance in the chromium-reducing method.

REFERENCES CITED

Andriessen, P.A.M., Helmens, K.F., Hooghiemsta, H., Riezebos, P.A., and van der Hammen, T., 1993, Absolute chronology of the Pliocene-Quaternary sediment sequence of the Bogotá area, Colombia: Quaternary Science Reviews, v. 12, p. 483–501, doi: 10.1016/0277-3791(93)90066-U.

Baker, L.A., Urban, N.R., Brezonic, P.L., and Sherman, L.A., 1989, Sulfur cycling in a seepage lake, *in* Saltzman, E., and Cooper, W., eds., Biogenic Sulfur in the Environment: Washington, D.C., American Chemical Society, p. 79–100.

Baker, P.A., Rigsby, C.A., Seltzer, G.O., Fritz, S.C., Lowenstein, T.K., Bacher, N.P., and Vellz, C., 2001, Tropical climate changes at millennial and orbital time scales on the Bolivian Altiplano: Nature, v. 409, p. 698–701, doi: 10.1038/35055524.

Bates, A.L., Spiker, E.C., Hatcher, P.G., Stout, S.A., and Weintraub, V.C., 1995, Sulfur geochemistry of organic-rich sediments from Mud Lake, Florida, USA: Chemical Geology, v. 121, p. 245–262, doi: 10.1016/0009-2541(94)00122-O.

Benning, L.G., Wilkin, R.T., and Barnes, H.L., 2000, Reaction pathways in the Fe-S system below 100 °C: Chemical Geology, v. 167, p. 25–51, doi: 10.1016/S0009-2541(99)00198-9.

Berger, A., Loutre, M.F., and Laskar, J., 1992, Stability of the astronomical frequencies over the Earth's history for paleoclimate studies: Science, v. 255, p. 560–566.

Berner, R.A., 1970, Sedimentary pyrite formation: American Journal of Science, v. 268, p. 1–8.

Berner, R.A., Baldwin, T., and Holdren, G.R., 1979, Authigenic iron sulfides as paleosalinity indicators: Journal of Sedimentary Petrology, v. 49, p. 1345–1350.

Brüchert, V., and Pratt, L.M., 1996, Contemporaneous early diagenetic formation of organic and inorganic sulfur in estuarine sediments from St. Andrew Bay, Florida, USA: Geochimica et Cosmochimica Acta, v. 60, p. 2325–2332, doi: 10.1016/0016-7037(96)00087-7.

Bush, M.B., 2002, On the interpretation of fossil *Poacea* pollen in the lowland humid neotropics: Palaeogeography, Palaeoclimatology, Palaeoecology, v. 177, p. 5–17, doi: 10.1016/S0031-0182(01)00348-0.

Canfield, D.E., 1989, Reactive iron in marine sediments: Geochimica et Cosmochimica Acta, v. 53, p. 619–632, doi: 10.1016/0016-7037(89)90005-7.

Canfield, D.E., and Des Marais, D.J., 1991, Aerobic sulfate reduction in microbial mats: Science Reprint Series, v. 251, p. 1471–1473.

Canfield, D.E., Raiswell, R., Westrich, J.T., Reaves, C.M., and Berner, R.A., 1986, The use of chromium reduction in the analyses of reduced inorganic sulfur in sediments and shales: Chemical Geology, v. 54, p. 149–155, doi: 10.1016/0009-2541(86)90078-1.

Colinvaux, P.A., De Oliveira, P.E., and Bush, M.B., 2000, Amazonian and neotropical plant communities on glacial time scales: The failure of the aridity and refuge hypotheses: Quaternary Science Reviews, v. 19, p. 141–169, doi: 10.1016/S0277-3791(99)00059-1.

Cook, R.B., 1984, Distribution of ferrous iron and sulfide in an anoxic hypolimnion: Canadian Journal of Fisheries and Aquatic Sciences, v. 41, p. 286–293.

Cook, R.B., and Kelly, C.A., 1992, Sulphur cycling and fluxes in temperate dimictic lakes, *in* Howard, R.W., Stewart, J.W.B., and Ivanov, M.V., eds., Sulphur Cycling on the Continents: Washington, John Wiley and Sons, p. 145–188.

Cook, R.B., Kreis, R.G., Kingston, J.C., Camburn, K.E., Norton, S.A., Mitchell, M.J., Fry, B., and Shane, L.C.K., 1990, Paleolimnology of McNearney Lake; an acidic lake in northern Michigan: Journal of Paleolimnology, v. 3, p. 13–34, doi: 10.1007/BF00209297.

David, M.B., and Mitchell, M.J., 1985, Sulfur constituents and cycling in water, seston, and sediment in an oligotrophic lake: Limnology and Oceanography, v. 30, p. 1196–1207.

de Menocal, P.B., Ruddiman, W.F., and Pokras, E.M., 1993, Influences of high- and low-latitude processes on the African terrestrial climate: Pleistocene eolian records from equatorial Atlantic Ocean Drilling Program Site 63: Paleoceanography, v. 8, p. 209–242.

Ferdelman, T.G., Church, T.M., and Luther, G.W., 1991, Sulfur enrichment of humic substances in a Delaware salt marsh sediment core: Geochimica et Cosmochimica Acta, v. 55, p. 979–988, doi: 10.1016/0016-7037(91)90156-Y.

Friese, K., Wendt-Potthoff, K., Zachman, D.W., Fauville, A., Mayer, B., and Veizer, J., 1998, Biogeochemistry of iron and sulfur in sediments of an acidic mining lake in Lusatia, Germany: Water, Air, and Soil Pollution, v. 108, p. 231–247, doi: 10.1023/A:1005195617195.

Haberle, S.G., and Maslin, M.A., 1999, Late Quaternary and climate change in the Amazon Basin based on a 50,000 year pollen record from the Amazon Fan, ODP Site 932: Quaternary Research, v. 51, p. 27–38, doi: 10.1006/qres.1998.2020.

Harris, S.E., and Mix, A.C., 1999, Pleistocene precipitation balance in the Amazon Basin recorded in deep sea sediments: Quaternary Research, v. 51, p. 14–26, doi: 10.1006/qres.1998.2008.

Hartgers, W.A., Lopez, J.F., Sinninghe Damsté, J.S., Reiss, C., Maxwell, J.R., and Grimalt, J.O., 1997, Sulphur-binding in recent environments. II. Speciation of sulphur and iron and implications for the occurrence of organo-sulphur compounds: Geochimica et Cosmochimica Acta, v. 61, p. 4769–4788, doi: 10.1016/S0016-7037(97)00279-2.

Haug, G.H., Hughen, K.A., Sigman, D.M., Peterson, L.C., and Rohl, U., 2001, Southward migration of the Intertropical Convergence Zone through the Holocene: Science, v. 293, p. 1304–1308, doi: 10.1126/science.1059725.

Hays, J.D., Imbrie, J., and Shackleton, N.J., 1976, Variations in the Earth's orbit: Pacemaker of the ice ages: Science, v. 194, p. 1121–1132.

Hodell, D.A., Curtis, J.H., Jones, G.A., Higuera-Gundy, A., Brenner, M., Binford, M.W., and Dorsey, K.T., 1991, Reconstruction of Caribbean climate change over the past 10,500 years: Nature, v. 352, p. 790–793, doi: 10.1038/352790a0.

Hooghiemstra, H., and Ran, E.T.H., 1994, Late and middle Pleistocene climatic change and forest development in Colombia: Pollen record Funza-II (2–158 m core interval): Palaeogeography, Palaeoclimatology, Palaeoecology, v. 109, p. 211–246, doi: 10.1016/0031-0182(94)90177-5.

Hooghiemstra, H., Melice, J.L., Berger, A., and Shackleton, N.J., 1993, Frequency spectra and paleoclimatic variability of the high-resolution 30–1450 ka Funza I pollen record (Eastern Cordillera, Colombia): Quaternary Science Reviews, v. 12, p. 141–156, doi: 10.1016/0277-3791(93)90013-C.

Jorgensen, B.B., 1977, The sulfur cycle of a coastal marine sediment (Limfjorden, Denmark): Limnology and Oceanography, v. 22, p. 814–832.

Jorgensen, B.B., 1990, The sulfur cycle of freshwater sediments: Role of thiosulfate: Limnology and Oceanography, v. 35, p. 1329–1342.

King, G.M., and Klug, M.J., 1982, Comparative aspects of sulfur mineralization in sediments of a eutrophic lake basin: Applied and Environmental Microbiology, v. 43, p. 1406–1412.

Kleeberg, A., 1997, Interactions between benthic phosphorous release and sulfur cycling in Lake Scharmutzelsee (Germany): Water, Air, and Soil Pollution, v. 99, p. 391–399, doi: 10.1023/A:1018383221469.

LaLonde, R.T., 1990, Polysulfide reactions in the formation of organosulfur compounds in the geosphere, *in* Orr, W.L., and White, C.M., eds., Geochemistry of Sulfur in Fossil Fuels: American Chemical Society Symposium Series, v. 429, p. 68–82.

Landers, D.H., and Mitchell, M.J., 1988, Incorporation of $^{35}SO_4$ into sediments of three New York Lakes: Hydrobiologia, v. 160, p. 85–95.

Laskar, J., Jouterl, F., and Boudin, F., 1993, Orbital, precessional, and insolation quantities for the Earth from –20 Myr to +10 Myr: Astronomy and Astrophysics, v. 270, p. 522–533.

Lovley, D.R., and Klug, M.J., 1986, Model for the distribution of sulfate reduction and mathanogenesis in freshwater sediments: Geochimica et Cosmochimica Acta, v. 50, p. 11–18, doi: 10.1016/0016-7037(86)90043-8.

Luther, G.W., 1991, Pyrite synthesis via polysulfide compounds: Geochimica et Cosmochimica Acta, v. 55, p. 2839–2849, doi: 10.1016/0016-7037(91)90449-F.

Lyons, T.W., 1997, Sulfur isotopic trends and pathway of iron sulfide formation in upper Holocene sediments of the anoxic Black Sea: Geochimica et Cosmochimica Acta, v. 61, p. 3367–3382, doi: 10.1016/S0016-7037(97)00174-9.

Marnette, E.C., Hordijk, C., Breemen, N.V., and Cappenberg, T.E., 1992, Sulfate reduction and S-oxidation in a moorland pool sediment: Biogeochemistry, v. 17, p. 123–143, doi: 10.1007/BF00002643.

Martin, L., Bertaux, J., Correge, T., Ledru, M.P., Mourguiart, P., Sideddine, A., Soubies, F., Wirrmann, D., Seuguio, K., and Turcq, B., 1997, Astronomical forcing of contrasting rainfall changes in tropical South America between 12,400 and 8800 cal yr B.P.: Quaternary Research, v. 47, p. 117–122, doi: 10.1006/qres.1996.1866.

Mayle, F.E., Burbridge, R., and Killeen, T.J., 2000, Millennial-scale dynamics of southern Amazonian rain forests: Science, v. 290, p. 2291–2294, doi: 10.1126/science.290.5500.2291.

Middelburg, J.J., 1991, Organic carbon, sulphur, and iron in recent semi-euxinic sediments of Kau Bay, Indonesia: Geochimica et Cosmochimica Acta, v. 55, p. 815–828, doi: 10.1016/0016-7037(91)90344-5.

Mitchell, M.J., Owen, J.S., and Schindler, S.C., 1990, Factors affecting incorporation of sulfur into lake sediments: Paleoecological implications: Journal of Paleolimnology, v. 4, p. 1–22, doi: 10.1007/BF00208295.

Mix, A.C., Ruddiman, W.F., and McIntyre, A., 1986, Late Quaternary paleoceanography of the tropical Atlantic, 1, Spatial variability of annual mean sea-surface temperatures, 0–20,000 years B.P.: Paleoceanography, v. 1, p. 43–66.

Molongoski, J.J., and Klug, M.J., 1980, Anaerobic metabolism of particulate organic matter in the sediments of a hypereuthrophic lake: Freshwater Biology, v. 10, p. 507–518.

Mommersteeg, H., 1998, Vegetation development and cyclic and abrupt climatic change during the late Quaternary; palynological evidence from the Eastern Cordillera of Colombia [Ph.D. thesis]: Amsterdam, University of Amsterdam, 191 p.

Mora, G., and Pratt, L.M., 2001, Isotopic evidence for cooler and drier conditions in the tropical Andes during the last glacial stage: Geology, v. 29, p. 519–522, doi: 10.1130/0091-7613(2001)029<0519:IEFCAD>2.0.CO;2.

Mora, G., Pratt, L.M., Boom, A., and Hooghiemstra, H., 2002, Biogeochemical characteristics of lacustrine sediments reflecting a changing alpine neotropical ecosystem during the Pleistocene: Quaternary Research, v. 58, p. 189–196, doi: 10.1006/qres.2002.2369.

Nriagu, J.O., and Soon, Y.K., 1985, Distribution and isotope composition of sulfur in lake sediments of northern Ontario: Geochimica et Cosmochimica Acta, v. 49, p. 823–834, doi: 10.1016/0016-7037(85)90175-9.

Paillard, D., Labeyrie, L., and Yiou, P., 1996, Macintosh program performs time-series analysis: Eos (Transactions, American Geophysical Union), v. 77, p. 379.

Peterson, L.C., Overpeck, J.T., Kipp, N.J., and Imbrie, J., 1991, A high-resolution late Quaternary record from the anoxic Cariaco basin, Venezuela: Paleoceanography, v. 6, p. 99–119.

Putschew, A., Scholz-Böttcher, B.M., and Rullkötter, J., 1996, Early diagenesis of organic matter and related sulphur incorporation on surface sediments of meromictic Lake Cadagno in the Swiss Alps: Organic Geochemistry, v. 25, p. 379–390, doi: 10.1016/S0146-6380(96)00143-X.

Pyzik, A.J., and Sommer, S.E., 1981, Sedimentary iron monosulfides; kinetics and mechanisms of formation: Geochimica et Cosmochimica Acta, v. 45, p. 687–698, doi: 10.1016/0016-7037(81)90042-9.

Rice, C.A., Tuttle, M.L., and Reynolds, R.L., 1993, The analysis of forms of sulfur in ancient sediments and sedimentary rocks: Comments and cautions: Chemical Geology, v. 107, p. 83–95, doi: 10.1016/0009-2541(93)90103-P.

Riehl, H., 1979, Climate and weather in the tropics: San Diego, Academic Press, 611 p.

Roden, E.E., and Tuttle, J.H., 1993, Inorganic sulfur turnover in oligohaline estuarine sediments: Biogeochemistry, v. 22, p. 81–105.

Sass, H., Cypionka, H., and Babenzien, H.D., 1997, Vertical distribution of sulfate-reducing bacteria at the oxic-anoxic interface in sediments of oligotrophic Lake Stechlin: FEMS Microbiology Ecology, v. 22, p. 245–255, doi: 10.1016/S0168-6496(96)00096-7.

Schouten, S., de Graaf, W., Sinninghe Damsté, J.S., van Driel, G.B., and de Leeuw, J.W., 1994, Laboratory simulation of natural sulphurization. II. Reaction of multifunctionalized lipids with inorganic polysulphides at low temperatures: Organic Geochemistry, v. 22, p. 825–834, doi: 10.1016/0146-6380(94)90142-2.

Seltzer, G.O., Rodbell, D., and Burns, S., 2000, Isotopic evidence for late Quaternary climatic change in tropical South America: Geology, v. 28, p. 35–38, doi: 10.1130/0091-7613(2000)028<0035:IEFLQC>2.3.CO;2.

Shackleton, N.J., Berger, A., and Peltier, W.R., 1990, An alternative astronomical calibration of the lower Pleistocene time scale based on ODP Site 677: Transactions of the Royal Society of Edinburgh: Earth Sciences, v. 81, p. 251–261.

Sinke, A.J.C., Cornelese, A.A., Cappenberg, T.E., and Zehnder, A.J.B., 1992, Seasonal variation in sulfate reduction and mathanogenesis in peaty sediments of eutrophic Lake Loosdrecht, The Netherlands: Biogeochemistry, v. 16, p. 43–61, doi: 10.1007/BF00024252.

Sinninghe Damsté, J.S., and de Leeuw, J.W., 1990, Analysis, structure and geochemical significance of organically bound sulphur in the geosphere: State of the art and future research: Organic Geochemistry, v. 16, p. 1077–1101, doi: 10.1016/0146-6380(90)90145-P.

Sinninghe Damsté, J.S., Kok, M.D., Köster, J., and Schouten, S., 1998, Sulfurised carbohydrates: An important sedimentary sink for organic carbon?: Earth and Planetary Science Letters, v. 164, p. 7–13, doi: 10.1016/S0012-821X(98)00234-9.

Smith, R.L., and Klug, M.J., 1981, Reduction of sulfur compounds in the sediments of a euthrophic lake basin: Applied and Environmental Microbiology, v. 41, p. 1230–1237.

Thomson, D.J., 1982, Spectrum estimation and harmonic analysis: Proceedings of the Institute of Electrical and Electronics Engineers, v. 70, p. 1055–1096.

Urban, N.R., and Monte, A.E., 2001, Sulfur burial and loss from the sediments of Little Rock Lake, Wisconsin: Canadian Journal of Fisheries and Aquatic Sciences, v. 58, p. 1347–1355, doi: 10.1139/cjfas-58-7-1347.

Vairavamurthy, A., and Mopper, K., 1987, Geochemical formation of organosulphur compounds (thiols) by addition of H_2S to sedimentary organic matter: Nature, v. 329, p. 623–625, doi: 10.1038/329623a0.

Vairavamurthy, A., Zhou, W., Eglinton, T., and Manowitz, B., 1994, Sulfonates; a novel class of organic sulfur compounds in marine sediments: Geochimica et Cosmochimica Acta, v. 58, p. 4681–4687, doi: 10.1016/0016-7037(94)90200-3.

van der Hammen, T., and Absy, M.L., 1994, Amazonia during the last glacial: Palaeogeography, Palaeoclimatology, Palaeoecology, v. 109, p. 247–261, doi: 10.1016/0031-0182(94)90178-3.

Wakeham, S.G., Sinninghe Damsté, J.S., Kohnen, M.E.L., and de Leeuw, J.W., 1995, Organic sulfur compounds formed during early diagenesis in Black Sea sediments: Geochimica et Cosmochimica Acta, v. 59, p. 521–533, doi: 10.1016/0016-7037(94)00361-O.

Werne, J., Hollander, D.J., Behrens, A., Schaeffer, P., Albrecht, P., and Sinninghe Damsté, J.S., 2000, Timing of early diagenetic sulfurization of organic matter: A precursor-product relationship in Holocene sediments of the anoxic Cariaco Basin, Venezuela: Geochimica et Cosmochimica Acta, v. 64, p. 1741–1751, doi: 10.1016/S0016-7037(99)00366-X.

Wetzel, R.G., 1972, The role of carbon in hard-water marl lakes, *in* Likens, G.E., ed., Nutrients and Eutrophication: The Limiting-Nutrient Controversy, Special Symposium of American Society of Limnology and Oceanography, v. 1, p. 84–97.

Wetzel, R.G., 1983, Limnology: Philadelphia, W.B. Saunders, 767 p.

Xia, K., Weesner, F., Bleam, W.F., Bloom, P.R., Skyllberg, U.L., and Helmke, P.A., 1998, XANES studies of oxidation states of sulfur in aquatic and soil humic substances: Soil Science Society of America Journal, v. 62, p. 1240–1246.

Manuscript Accepted by the Society 19 April 2005

Printed in the USA

Geological Society of America
Special Paper 395
2005

Use of hydrogen isotope variations in speleothem fluid inclusions as an independent measure of paleoclimate

Feride Serefiddin
Department of Chemistry and Chemical Biology, Rutgers University Piscataway, New Jersey 08854, USA

Henry Schwarcz
Derek Ford
School of Geography and Earth Sciences, McMaster University, Hamilton, Ontario L8S 4M1, Canada

ABSTRACT

Fluid inclusions trapped in speleothems represent samples of the drip water (dw) from which the speleothems grew. It is uncertain whether the original $^{18}O/^{16}O$ ratios may have subsequently been altered due to exchange with the calcite (ct). We can use their deuterium/hydrogen (D/H) ratios to reconstruct $\delta^{18}O$ of the drip water using the local meteoric water line (MWL) giving the relationship between $\delta^{18}O$ and δD. We can thus calculate paleotemperatures from the calcite-water isotopic fractionation if the MWL relationship is known. A Holocene speleothem from Santa Cruz, California, displays a large variation in δD, possibly due to changing amount effect, and yields temperatures at 4–3.5 k.y. B.P. –7 °C lower than modern. Two coeval, mid-Wisconsin speleothems deposited tens of meters apart in Reed's Cave, South Dakota, exhibit different $\delta^{18}O_{ct}$ values but, when combined with $\delta^{18}O_{dw}$, we obtain similar growth temperatures. Thus speleothems may display divergent $\delta^{18}O_{ct}$ records, but still record the same long-term climate variations.

Keywords: speleothems, hydrogen isotopes, fluid inclusions, paleotemperatures.

INTRODUCTION

The application of hydrogen isotope ratio (δD) measurements to paleoclimate studies relies on the accurate and precise measurement of isotopic composition in fluid inclusions. The ratio of stable isotopes of hydrogen in speleothem fluid inclusions can be used to reconstruct changes in precipitation and calculate paleotemperatures as independent measures of climate change. The application of uranium-series disequilibrium dating gives these proxies high-precision chronological control (Schwarcz, 1986; Dorale, 2000). The hydrogen isotope composition of cave drip waters has been shown to be a good estimate of surface precipitation (Yonge et al., 1985). Therefore the trapped inclusions are preserving the original isotope signal of precipitation. Fractionation during extraction of these fluid inclusion waters is the only significant complicating factor observed from previous studies.

Previous Fluid Inclusion Research with Speleothems

During crystal growth speleothems can trap and preserve some of the original seepage water as fluid inclusions, providing a source of information on the past isotopic composition of precipitation. Fluid inclusions often comprise as much as 0.1 to >1.0 wt% of a speleothem (Schwarcz et al., 1976). Pioneering research in the 1970s on speleothems found that δD of fluid inclusions was equivalent to surface precipitation waters (Schwarcz et al., 1976; Harmon et al., 1979; Harmon and Schwarcz, 1981; Schwarcz and Yonge, 1983; Yonge et al., 1985). Rozanski and Dulinski (1987) found similar results but their calcite sample size was too

Serefiddin, F., Schwarcz, H., and Ford, D., 2005, Use of hydrogen isotope variations in speleothem fluid inclusions as an independent measure of paleoclimate, *in* Mora, G., and Surge, D., eds., Isotopic and elemental tracers of Cenozoic climate change: Geological Society of America Special Paper 395, p. 43–53, doi: 10.1130/2005.2395(05). For permission to copy, contact editing@geosociety.org.

large to give good high-resolution data. Later researchers found a 20–30‰ offset in δD values when the water was removed by heating at temperatures over 800 °C versus extraction at lower temperatures (Yonge, 1982; Goede et al., 1986, 1990; Matthews et al., 2000; McGarry et al., 2002). This offset was believed to be a result of fractionation occurring during the heating and extraction process. This offset could also be due to an additional component of highly fractionated water that is bound to the calcite lattice (Yonge, 1982). Thermal desorption of released water from the crushed calcite at 150 °C gives almost complete recovery of extracted water and showed negligible offset, so the water released in the high temperature experiments may be related to additional, structurally bound water (Dennis et al., 2001). Recent work in the McMaster University laboratory, in collaboration with G. Rossman and C. Verdel, Caltech, shows that nanometer size inclusions may be present that can remain in the calcite, even after repeated crushing and heating of the sample material.

Paleoclimate Reconstruction using Fluid Inclusions

When calcite forms in oxygen isotopic equilibrium with water, we can calculate the temperature of formation from the isotopic fractionation, $\alpha_{c\text{-}w}$, between calcite and water, where $\alpha_{c\text{-}w} = (^{18}O/^{16}O)_{calcite}/(^{18}O/^{16}O)_{water}$. The calculation of paleotemperatures using the O'Neil et al. (1969) calcite-water fractionation equation

$$1000 \ln \alpha_{c\text{-}w} = 2.78 \times 10^{-6}\, T^{-2} - 2.89 \quad (1)$$

requires knowledge of the $\delta^{18}O$ of the drip water from which the speleothem was precipitated. We assume that this is equal to the initial $\delta^{18}O$ of fluid inclusions (fi) trapped in the speleothem. Although these fluid inclusions can be analyzed for both $\delta^{18}O$ and δD, $\delta^{18}O$ is not used to calculate paleotemperatures because the oxygen isotopes of the calcite and the water may have exchanged following deposition of the speleothem. Instead we have calculated the initial $\delta^{18}O$ value of the trapped water from its measured δD. We use the relationship between $\delta^{18}O$ and δD observed in local meteoric waters (meteoric water line, MWL) which is generally represented as $\delta D = 8\ \delta^{18}O + \delta_o$ (deuterium excess, D-excess, $\delta_o = 8\ \delta^{18}O - \delta D$). The absence of hydrogen in the surrounding calcite ensures that no exchange is taking place, but does require the meteoric water relationship for calculations.

Dennis et al. (2001) showed that the $\delta^{18}O$-δD relation from inclusions extracted from a Holocene speleothem in Great Britain lay on the present-day meteoric water line (δ_o = 10‰). McDermott et al. (2005) have given various arguments suggesting that oxygen-isotopic exchange is negligible in Pleistocene fluid inclusions. Schwarcz and Yonge (1983) found, however, that $\delta^{18}O$-δD data for some Pleistocene speleothems could not be accounted for by changes in δ_o. In the present study we have calculated the initial $\delta^{18}O$ value of the trapped water from its measured δD, using the appropriate meteoric water line for the period in which the sample grew.

The modern meteoric water line relationship is defined to be $\delta D = 8\ \delta^{18}O + 10$ (Craig, 1961; Dansgaard, 1964). This is the global average for the relationship between deuterium and oxygen isotopes in precipitation. The slope and D-excess can change with location, seasonally and over time (Rozanski et al., 1992; Nativ and Riggio, 1990; Merlivat and Jouzel, 1969). This relationship is used to calculate $\delta^{18}O$ values from the measured δD in fluid inclusions. The major obstacle to the calculation of reliable paleotemperatures is determining the MWL relationship in the past. Modern studies of the MWL show variability in this relationship between summer and winter seasons due to the amount effect and proximity to the oceans (Rozanski et al., 1993). Nativ and Riggio (1990) found that the MWL in eastern New Mexico had a winter average of $\delta D = 7.7\ \delta^{18}O + 13.2$ and a summer average of $\delta D = 6.9\ \delta^{18}O - 0.53$. Longinelli and Selmo (2003) found a spatial range in values for the slope of the MWL in Italy from 5.7 to 8.9 and in the deuterium excess from 9.2 to 19.1. Genty et al. (2002) measured δD in large fluid-filled voids in a speleothem and calculated a −1.3 °C temperature difference from modern for a growth layer in a French speleothem dated at 100 ka. The $\delta^{18}O$ of fluid inclusion water was calculated using the measured δD and local meteoric water line relationship $\delta D = 8\delta^{18}O + 7.7$ (Genty et al., 2002). The growth layer grew during marine isotope substage 5c, a period in the last interglacial that appears to have been slightly cooler than present (Martinson et al., 1987; Jouzel et al., 1987). Late Holocene records from Great Britain show isotopic composition of fluid inclusion waters that are not significantly different from modern drips (Dennis et al., 2001).

In the present study we have made use of local MWLs obtained from analyses of drip waters and local precipitation in order to reconstruct $\delta^{18}O$ of fluid inclusions from δD values. As will be seen, it was not possible to infer specific changes that may have taken place in the MWL in the past. Lack of this information would contribute to some uncertainty in the reconstruction of paleotemperatures.

METHODS

Fluid inclusion water was extracted using equipment originally designed at the University of East Anglia Isotope Lab (Dennis et al., 2001). The apparatus consists of a crushing cell that connects to a vacuum line. The crushing cell is comprised of a tower and crushing chamber, with a piston controlled by an electromagnet (Fig. 1). The entire crushing apparatus is placed under vacuum and fitted with heaters for the removal of water that may adsorb to the walls of the chamber after crushing. Heating tape is wrapped around the vacuum line to improve movement of water along the line to the cold trap.

Slices of calcite were taken along growth layers with a maximum thickness of 5 mm using a low speed Isomet diamond wafer saw. At least two samples were cut from each growth layer for repeat analyses. The age of samples was determined from an age model developed from a series of uranium-series dates. A more detailed description of dating locations and technique can be

Figure 1. Crushing cell for extraction of fluid inclusion water.

Figure 2. Vacuum line for collection of fluid inclusion water.

found in a previous paper (Serefiddin et al., 2004). A portion of the slice weighing ~0.5–1.0 g was loaded into the cell and evacuated. After the sample had been heated at 100 °C for 15 min, it was gently crushed for 15 min by repetitive motion of the piston. During this time, the crushing cell was closed off from the high vacuum pumps and opened to the U-trap (Fig. 2). The U-trap was cooled with liquid nitrogen to collect the water and CO_2 released during crushing. After crushing, the cell was heated to 150 °C to remove all adsorbed water. The liquid nitrogen was then replaced with dry ice–ethanol slurry to release CO_2, which was pumped away. The cold trap was warmed to room temperature and water was trapped in a Pyrex tube containing 50 mg of zinc shavings ("Indiana Zn"). The water was then reacted with the zinc at 500 °C for 1 h to produce hydrogen gas, which was analyzed on a SIRA II mass spectrometer against H_2 from a laboratory standard (DTAP). The absolute δD values were calculated using the VSMOW and VSLAP reference waters for calibration. The precision of measurements of the standard is 0.11‰.

The crushed calcite was then reserved for analysis of $\delta^{18}O_{ct}$. The entire crushed sample of calcite was ground in an agate mortar and homogenized. A 100 μg sample was analyzed to measure $\delta^{18}O_{ct}$ of the growth layer. It is assumed that both the $\delta^{18}O_{ct}$ and the δD_{fi} represent an average over the time period during which the sample grew.

The volume of water in each calcite sample was estimated from the intensity of the major beam (^2H signal). Capillaries containing known masses of water were used to create a calibration curve. Twelve samples gave a beam intensity of less than 5.0×10^{-10}, which indicated less than 0.5 μL of water; these data were not used because they were most likely fractionated during extraction (Dennis et al., 2001).

Samples

Samples were prepared for fluid inclusion analysis to evaluate δD and temperature variability during the Wisconsin glacial stage from Reed's Cave, South Dakota, and during the Holocene at Vanishing River Cave, California (Fig. 3). In an attempt to produce high-resolution records, sample size was reduced to ~1 g or less.

Speleothems RC2 and RC20 are two samples from Reed's Cave, South Dakota, that partly grew at the same time, from 62 to 49 k.y. B.P. These speleothems have partly divergent $\delta^{18}O_{ct}$ records, but each appears to be recording climate, with local effects superimposed on the long-term records (Serefiddin et al., 2004). The analysis of the δD and paleotemperature calculations allows us to test whether similar temperatures are being recorded by these two speleothems. Time resolution for each measurement ranges from 200 yr during fast growth to 6300 yr during slow growth for sample RC2. The faster growing speleothem RC20 has a time resolution ranging from 50 yr to 690 yr. Extra care was taken to ensure that the samples were from the central growth axis of the speleothem, where the growth layers are thickest and there is the greatest likelihood of equilibrium conditions.

A modern speleothem from Vanishing River Cave, Santa Cruz, was analyzed to investigate temperature gradients and the

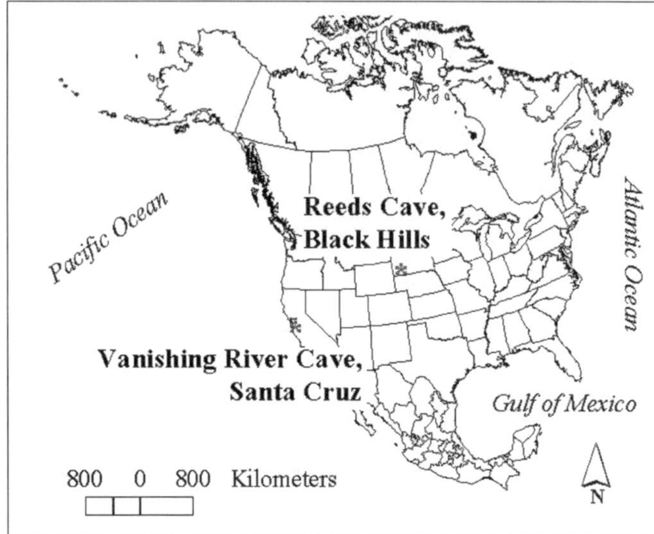

Figure 3. Site map showing location of Reed's Cave in the Black Hills of South Dakota and Vanishing River Cave in Santa Cruz on central California Coast.

TABLE 1. RESULTS OF STANDARDS ANALYSES

Sample ID	δD	precision	m/l	Measured water vol./μL
Laboratory standard water				
DTAP*	–57.6	0.1		
Standard water filled capillary				
CAP-2	–54.2	0.1	n/a	2.0
Standard water filled capillaries run with Iceland Spar				
MEXIS-1	–56.6	0.1	nr	2.0
MEXIS-4	–66.9	0.1	nr	1.0
MEXIS-5	–61.4	0.1	0.0	1.0
MEXIS-7	–58.1	0.1	0.0	0.5
MEXIS-8	–98.9	0.4	0.0	0.5
MEXIS-9	–49.1	0.1	0.0	2.0
MEXIS-10	–53.7	0.1	0.0	1.0
MEXIS-11	–66.7	0.1	nr	
MEXIS-14	–106.9	0.1	dnr	1.0
MEXIS-15	–68.2	0.1	0.0	1.0
MEXIS-17	–7.2	0.1	dnr	0.5
MEXIS-18	–75.7	0.1	0.0	0.5
MEXIS-20	–77.1	0.1	dnr	1.0
MEXIS-21	–59.3	0.2	0.0	1.0
MEXIS average†	–60.1	0.8		

Note: nr—not recorded; dnr—did not report.
*mean values for 60 samples
†n = 8

precipitation patterns in western North America. Speleothem VR-1 from Santa Cruz grew from 8.5 to 0.5 k.y. B.P. at sea level in a marine Mediterranean climate zone. The translucent, brown calcite has dissolution pockets containing clay detritus. These areas were avoided when preparing the samples. The oxygen isotope record shows very little variation (Holden et al., 2001, personal commun.) The sample size averaged 5 mm thick and averaged 300 yr of growth. A minimum of two samples was taken from each growth layer, with a third cut if there were enough samples available.

RESULTS

Tests of the Procedure

Calibration of the crushing cell and vacuum line was done by measurements of an internal laboratory (DTAP) standard water in glass capillaries alone and with sparry calcite following the procedure of Dennis et al. (2001). The reproducibility of samples was tested by analyzing replicates of the laboratory standard water, DTAP, in capillaries, together with 500–1000 mg pieces of Mexican calcite spar (MEXIS). Fourteen MEXIS samples were prepared but seven samples (MEXIS-1, 4, 8, 11, 14, 17, and 20) did not produce adequately large transducer readings or readings that indicated they were probably fractionated during transfer due to incomplete recovery or desorption (Table 1). The seven remaining samples gave an average δD value of –60.1 ± 8.2‰ with a range of –49.1 to –77.1‰. The average δD agrees with the δD value for laboratory standard DTAP of –57.6‰. The slightly depleted average value of –60.1‰ may result from fractionation effects from incomplete desorption of water from the calcite powder.

As another test of reproducibility of the method, sixteen growth layers of speleothems from Reed's Cave were analyzed in replicate, by analysis of slices taken as close as possible to the same level in the stalagmite. From the eleven layers, which had sufficient water for isotopic analyses, four of these results agreed within the average standard deviation of 3.3‰ (Table 2). Overall, the difference between replicate analyses ranges from 1 to 27‰. Two possible reasons for this are: (a) incomplete recovery of water, resulting in isotopic fractionation; and (b) nonequivalence of the supposed replicates such that different growth periods are being averaged. Figure 4 shows the cross section of samples RC2 and RC20, where it is clear that a sample taken

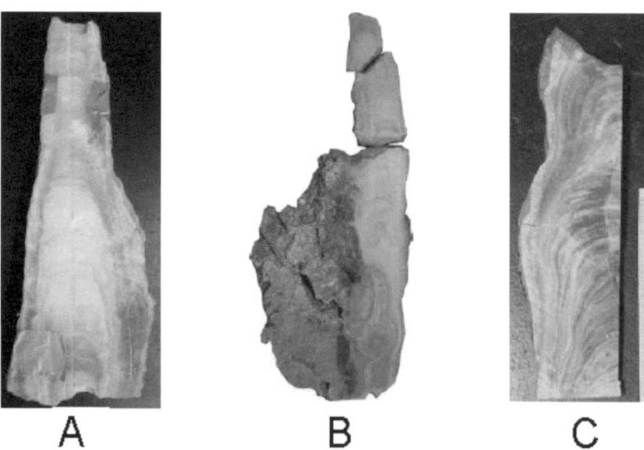

Figure 4. Speleothems studied here. (A) Reed's Cave 20000. (B) Reed's Cave 99902. (C) Santa Cruz VR-1.

TABLE 2. FLUID INCLUSION SAMPLE RESULTS

Sample ID	δD (‰ VSMOW)	Precision	Beam intensity*	Calculated volume (μL*)	Sample weight (g)	$H_2O/CaCO_3$ (mg/g)*
RC20_0.6.7a	–150.3	0.5	n/a		1.0	
RC20_0-6.7	–120.5	0.2	4.1E-10	0.4	1.0	0.4
RC20_0-6.7b	–105.6	0.4	3.3E-10	0.4	0.4	0.8
RC20-174.25	–125.0	0.2	6.4E-10	0.5	0.6	0.9
RC20-174.25A	–117.6	0.2	6.5E-10	0.5	0.9	0.6
RC20-183.35	–90.6	0.0	1.1E-09	0.8	0.9	0.8
RC20-183.35A	–94.2	0.1	1.4E-09	0.9	1.2	0.8
RC20-183.35B	–35.1	0.1	1.0E-09	0.7	0.7	1.0
RC20-33a	–82.7	0.2	2.0E-09			
RC20-34-1	–104.0	0.0	n/a		0.4	
RC20-34-2	–97.8	0.1	7.1E-10	0.6	0.4	1.4
RC20-35	–65.0	0.1	3.1E-09			
RC20-35a	–120.5	0.1	6.5E-10			
RC20-35b	–82.9	0.1	3.0E-09			
RC20-40	–64.9	0.1	8.2E-10	0.6	1.1	0.6
RC20-40a	–99.0	0.0	7.5E-10	0.6	1.1	0.6
RC20-40B	–123.8	0.1	8.6E-10	0.6	0.5	1.2
RC20-41a	–122.8	0.0	5.7E-10	0.5	0.6	0.8
RC20-41c	–121.9	0.1	5.4E-10	0.5	0.8	0.6
RC20-93.35	–114.4	0.1	3.5E-09	2.1	1.2	1.7
RC20-93.35A	–73.5	0.2	4.2E-09	2.4	1.2	2.0
RC20-93.35B	–116.8	0.1	9.1E-10	0.7	1.5	0.5
RC2-1	–122.1	0.1	8.9E-10	0.7	0.3	2.0
RC2-10OUT	–93.6	0.1	5.8E-09			
RC2-2&3IN	–104.1	0.1	1.0E-09	0.7	0.6	1.3
RC2-34OUT	–74.8	0.1	1.4E-09			
RC2-35	–93.7	0.1	5.7E-10			
RC2-35a	–73.7	0.2	5.8E-10			
RC2-4&5IN	–91.6	0.0	2.0E-09	1.3	0.4	3.0
RC2-4&5OUT	–82.7	0.2	8.5E-10	0.6	0.4	1.5
RC2-6&7OUT	–100.2	0.1	9.5E-10	0.7	0.3	2.1
RC2-9I	–39.9	0.1	3.0E-09	1.8	0.5	3.3
RC2-9OUT	–66.9	0.1	3.0E-09	1.8	0.4	4.0
RC2-OL-1	–129.4	0.1	2.9E-10	0.3	0.3	1.0
VR-1-13	–68.7	0.1	8.6E-10	0.6	1.1	0.6
VR-1-13a	–29.3	0.1	3.3E-09	1.9	0.9	2.3
VR-1-2	–22.7	0.1	7.8E-10	0.6	0.9	0.7
VR-1-24	–29.7	0.1	2.4E-09	1.4		
VR-1-25a	–36.5	0.1	8.0E-10	0.6	1.0	0.6
VR-1-26-1	–55.4	0.1	6.2E-10	0.5	1.1	0.5
VR-1-26-2	–46.9	0.1	6.3E-10	0.5	1.0	0.5
VR-1-27	–44.1	0.1	1.1E-09	0.8	1.0	0.8
VR-1-27a	–40.9	0.1	2.3E-09	1.4	0.7	2.0
VR-1-27b	–73.3	0.3	4.1E-10	0.4	0.4	0.9
VR-1-3	–55.4	0.1	4.5E-10	0.4	0.6	0.7
VR-1-4	–54.8	0.3	3.8E-10	0.4	0.8	0.5
VR-1-5a	–51.7	0.1	5.7E-10	0.5	0.7	0.7

Note: VSMOW—Vienna standard mean ocean water.
*All missing data results from insufficient sample size; sample was too small to run or too small to produce δD value.

near the outer perimeter of the stalagmite may contain thousands more years of growth than a sample from the main growth axis. Although care was taken to sample from the central growth axis, it is likely that the outer edges of some pieces sampled additional growth layers.

To determine if leaks of atmospheric water vapor would contaminate samples, we attempted to collect water from the line without bringing the crushing cell and Pyrex line down to a hard vacuum. Three attempts failed to result in a measurable amount of water. We conclude that any small leaks had no effect on the isotopic values of collected water.

Erroneous δD can result if the proper zinc/water ratio is not used: the recommended ratio is of 50 mg Zn to 1 μL of water. We assume that the maximum water content of the speleothem is 0.1 wt%, and with a yield close to 100%, the sample sizes averaging 500 mg to 1 g will give us extracted water volumes from 0.5 to 1.0 μL. Capillaries filled with 0.5 μL and 1.0 μL of DTAP water were run with 40 mg, 50 mg, and 60 mg of zinc. These quantities of zinc gave reproducible and accurate results using standards of DTAP as small as 0.5 μl in capillaries with Iceland spar.

Analyses of Speleothems

Sample weights ranged from 0.27 to 1.48 g. The number and volume of fluid inclusions varied between the growth layers so the wt% of water varied somewhat with sample size (Table 2). The average wt% of all samples analyzed is 0.12 ± 0.09. The results of δD analyses for the crushed calcite samples include only the samples that were large enough to run on the mass spectrometer. Typically, samples less that 0.5 μl in size were too small for isotopic analysis. Of the 78 samples that were crushed and transferred to the zinc tubes, only 62 had enough water to collect δD data.

Reed's Cave, South Dakota

Cave drip waters were sampled to determine spatial variation throughout the cave and whether there was a distinct seasonal signal in the isotopic composition of these waters. Oxygen and hydrogen isotope measurements were also used to estimate a modern MWL for Reed's Cave. Results from oxygen and hydrogen analyses show an average of −97.3‰ for the δD values, −12.9‰ for the $\delta^{18}O$ values and a variation of up to 36‰ in the δD values and 5‰ in the $\delta^{18}O$ values for drips (Table 3). Samples were taken in April and August of 2001, as well as in June 2000 and display variation well within the seasonal range expected for mid-continental precipitation (Rozanski et al., 1993).

Data for 211 samples of modern meteoric water from the Black Hills are given in graphical form by Driscoll et al. (2002). The data are shown on Figure 5 where we have fitted a least-squares line to them and obtained the following MWL: $\delta D = 7.3\ \delta^{18}O - 3.4$. Both the slope and intercept of this line are strikingly different from the global MWL, and suggests that rain and snow falling in this district is derived from water vapor that has been recycled by prior continental evaporative cycling. Figure 5 also shows data for drip waters from Reed's Cave. These generally agree with Driscoll et al.'s data except for one point, which appears to be evaporatively enriched and was collected from cave popcorn, an evaporative speleothem feature.

Forty-five samples from the two speleothems were crushed. Of these, 34 gave δD values, while 11 failed due to insufficient quantities of water. The average δD for the 34 samples is −97.1‰ with a minimum value of −150.3‰ and a maximum value of −35.1‰ (Table 2). $\delta^{18}O$ values were calculated using the MWL from Driscoll et al. (2002). Five samples with volumes less than 0.5 μL may have been fractionated during transfer of water from crushing cell to the zinc-filled Pyrex tube.

Santa Cruz, California

The 13 samples from the Santa Cruz speleothem give a mean δD of −46.9‰, with a range from −73.3‰ to −22.7‰ (Table 2). The mean value for modern cave drips is −37.1 ± 1.2‰ from five repeat analyses of drips from one site. We would expect to see a similar average throughout the Holocene period. The full range of 50.6‰ for δD values from fluid inclusions is much higher than expected from temperature effects alone. $\delta^{18}O$ values were calculated using the global MWL and estimated local MWL. Two to three samples were run for each slice due to the larger diameter central growth axis for this speleothem (Table 2, Fig. 4C). Two of the replicates agreed well, with differences of 1‰ and 9‰.

TABLE 3. RESULTS OF DRIPWATER ANALYSES FROM REED'S CAVE AND VANISHING RIVER CAVE

Sample ID	δD (‰) VSMOW	$\delta^{18}O$ (‰) VSMOW
Reed's Cave		
A1	−94.4	−11.9
A13	−99.8	−12.4
A34	−90.0	−12.4
A35	n/a	−12.2
A36	−81.1	−11.1
A45	−90.0	−12.2
A55	−103.2	−13.9
A63	−91.7	−12.1
A95	−94.1	−13.0
CCTR_82601	−90.1	−12.4
D65	−96.6	−12.7
NF-3_82601	−75.2	−10.3
NF-6	−93.5	−12.1
NF-12	−99.9	−13.3
NF-15	−117.0	−15.3
NF-18	−100.7	−13.6
NF-20	−105.5	−13.9
NF-20B	−102.2	−13.6
NF-20H1	−101.1	−13.6
NF-20H2	−102.2	−13.7
NF-33	n/a	−14.4
NFM-2	−108.7	−14.0
NFP-1	−98.9	−13.3
NFP-2	−99.5	−13.4
NFR-2	−96.8	−11.6
OM-20-3	n/a	−11.7
OM-20-4	−84.6	−11.9
RC-1	−88.8	−12.8
RC-2	−95.4	−12.2
RC-3	−90.2	−11.7
RC-5	−92.6	−12.3
RC-6	−90.0	−11.2
RC-7	−93.8	−12.3
RC-8	−84.7	−11.0
RC-9	−92.4	−11.9
RC-10	−95.4	−8.7
RC-13	−87.8	−11.1
SC-1	−101.7	−13.3
SC-3	−98.5	−12.8
TWC_82601	−111.4	−15.1
*Santa Cruz**		
VL-0-1	−38.2	−5.4
VL-0-1a	−36.5	
VL-0-1b	−35.2	
VL-0-1c	−38.5	
VL-0-1d	−35.8	
VL-0-1e	−34.3	
VL-0-1f	−32.9	

*Only one oxygen analysis.

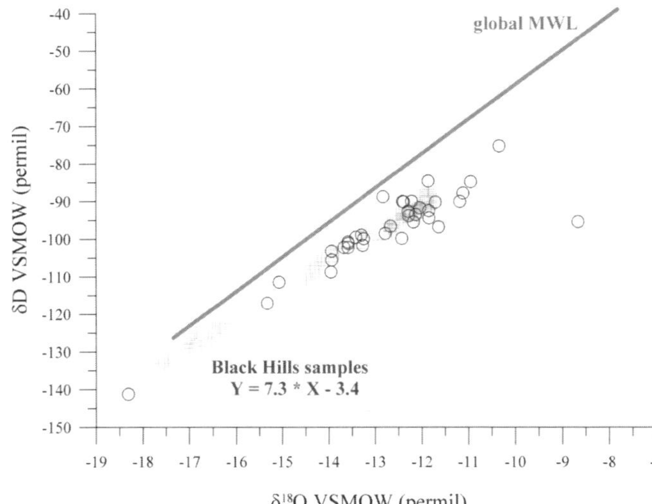

Figure 5. δD and $\delta^{18}O$ values of modern drip waters from Reed's Cave (O) and Black Hills surface and aquifer samples (+) from Driscoll et al. (2002), plotted with global meteoric water line (MWL). VSMOW—Vienna standard mean ocean water.

The third gave a difference of 40‰, which may be attributed to the problems described in the previous section. It reemphasizes the requirements for careful sample cutting along the main axis of the growth layers.

Variation in δD

The δD results clearly show that extraction and accurate measurement of fluid inclusion water is very difficult. Preliminary results give some reproducible data and result in calculated temperature shifts that can be correlated with other proxy records. Analysis of small samples and presence of bound water not released by crushing may also lead to unaccountable errors in δD values.

The average δD value for sample RC20 is consistently lower than for sample RC2, even during coeval periods of growth. A similar difference in average $\delta^{18}O_{ct}$ values for these two samples attributed this difference to seasonality of recharge of drip waters (Serefiddin et al., 2004). Sample RC20 was possibly formed from drips biased to winter precipitation. This could also explain the lower δD values in this sample. The range of δD values for sample RC20 is −65.0 to −150.3‰, lower than the range of δD values for sample RC2 of −39.9 to −129.4‰. There appears to be a slight enrichment of 10‰ in sample RC20 between 59 and 55 k.y. B.P. This could indicate a period of changing atmospheric circulation or temperature increase. There is a similar enrichment of δD values in sample RC2 beginning at ca. 57 k.y. B.P. There are comparable offsets in the $\delta^{18}O_{ct}$ and δD values for these two speleothems over this interval, so that the corresponding temperatures are in approximate agreement (see below). There is also evidence of warming in the $\delta^{13}C$ record at Crevice Cave in Missouri at this time (Dorale et al., 1998). The chronology for the change in RC2 is less certain due to weaker age control and lower resolution resulting from extremely slow growth rates over this interval.

Isotopic data from sample VR-1 in Vanishing River Cave, located on the central California Coast, could provide proxy data for climate change on the eastern Pacific margin. The δD values ranging from −23.3 to −73.3‰ are within expected values for δD measured from precipitation samples (International Atomic Energy Agency, 2001). The modern average δD from active drip sites in Vanishing River cave is –37.0‰. The time period represented from the fluid inclusion data is from 7.7 to 0.5 k.y. B.P. (Fig. 6). If the temperature dependent variation of δD in precipitation is the main control on the isotopic composition of these waters, these results can be interpreted as periods of temperatures colder and warmer than present. However, it is also possible that some of this variation is due to the "amount effect" (Rozanski et al., 1993). The "amount effect" is typically observed at sites between 20° N and 20° S, but has been reported as far north as Israel (23 °N; Bar-Matthews et al., 2000). If the amount of precipitation is controlling the δD values of these waters, the more negative values represent heavier rainfall periods and thus there are periods of more and less rainfall than modern during the Holocene. The δD value of –23‰ at 7.7 k.y. B.P. would indicate a period of decreased rainfall or warmer temperatures or a combination of these effects. This is consistent with the current climate regime in central and southern California that produces most of its rainfall during the winter.

The δD values dip below modern average δD at 3.7–4.0 k.y. B.P. with an average δD value of −45‰, indicating a period of cooler temperatures and/or increased rainfall. Increased rainfall from a strengthening of El Niño events is seen in modern coral records from the past 50 yr in the western Pacific (Gagan et al., 1998). Marine sediment records from the Santa Barbara Basin

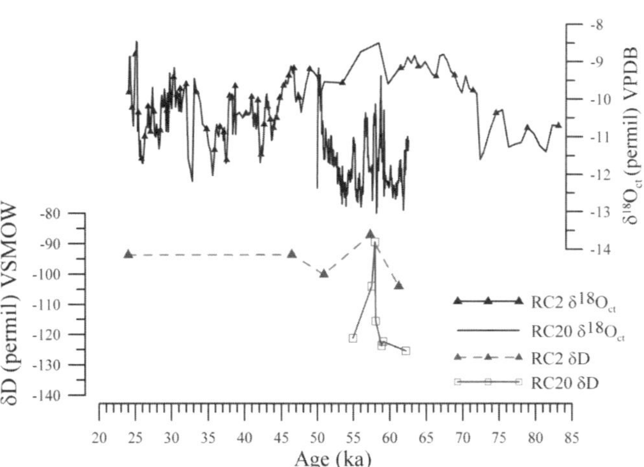

Figure 6. Graph showing $\delta^{18}O_{ct}$ values of Reed's Cave samples 99902 and 20000. Lower graph shows δD of measured fluid inclusions. Note how $\delta^{18}O_{ct}$ records are very different but the δD records are more coherent. VPDB—Vienna Peedee belemnite; VSMOW—Vienna standard mean ocean water.

(Hendy and Kennett, 1999) also support cooler temperatures during the 3.7–4.0 k.y. B.P. interval. On the other hand, pollen and lake-level records in the Sierra Nevada record decreases in effective moisture after 6 k.y. B.P. (Anderson, 1980; Thompson et al., 1993).

Paleotemperatures

Temperatures were calculated using the estimated $\delta^{18}O$ of fluid inclusions ($\delta^{18}O_{fi}$) and measured $\delta^{18}O_{ct}$ from the powders collected and homogenized after crushing. We calculated $\delta^{18}O$ of the fluid inclusions for Reed's Cave using the modern MWL obtained by Driscoll et al. (2002) and used these values to calculate temperatures as shown in Table 4. Some extracted waters yielded temperatures that are less than zero, while others yield T > 8 °C, the modern cave temperature. It is unlikely that cave temperatures during the cooler MIS 3 interglacial could have been greater than modern temperatures. These unrealistic temperatures may represent fluid samples with seasonal bias or contamination by strongly bound and fractionated lattice waters. They were omitted from the paleotemperature graphs for this reason. Replicate analyses along a selected sample layer can confirm that the resulting data represents accurate measurements and possibly true temperatures, not just randomly distributed values along the temperature range. Replicate analyses of sample layer 41 on RC20 give temperatures in agreement well within analytical error (Table 4).

The MWL relationship for the modern central California Coast is not well constrained, so two separate values are used as well. A lower δ_0 is seen in coastal and marine International Atomic Energy Agency sampling sites and is the basis for our value of $\delta_0 = 4$ (Rozanski et al., 1993; Dansgaard, 1964). Table 5 shows the temperatures calculated for the reduced δ_0 and also for the modern global MWL value of $\delta_0 = 10$. Using the same criteria

TABLE 4. RESULTS FROM PALEOTEMPERATURE CALCULATIONS FOR REED'S CAVE SPELEOTHEMS RC2 AND RC20

Sample ID[a]	Age (ka)	δD_{fi} (‰) VSMOW	$\delta^{18}O_{fi}$ (‰) (slope=7.3, d_0=-3.4)[b]	$\delta^{18}O$ (‰) VPDB	T[b] (calculated)
RC2-1	**61.9**	**−122.1**	**−16.3**	**−9.0**	**−8.2**
RC2-2&3IN	61.2	−104.1	−13.8	−9.3	−1.6
RC2-4&5IN	**57.3**	**−91.6**	**−12.1**	**−9.1**	**2.9**
RC2-4&5OUT	**57.3**	**−82.7**	**−10.9**	**−9.1**	**6.9**
RC2-4&5avg	57.3	−87.2	−11.5	−9.1	4.8
RC2-6&7OUT	50.9	−100.2	−13.3	−9.5	0.5
RC2-9I	**48.3**	**−39.9**	**−5.0**	**−9.7**	**34.4**
RC2-9OUT	**48.3**	**−66.9**	**−8.7**	**−9.7**	**17.4**
RC2-10OUT	46.4	−93.6	−12.4	−9.7	4.1
RC2-34OUT	**24.5**	**−74.8**	**−9.8**	**−10.0**	**14.4**
RC2-35	24.0	−93.7	−12.4	−9.2	2.2
RC2-35a	**24.0**	**−73.7**	**−9.6**	**−9.2**	**11.7**
RC2-OL-1	**24.0**	**−129.4**	**−17.3**	**−9.2**	**−9.8**
RC20-174.25	**54.9**	**−125.0**	**−16.7**	**−11.5**	**−2.8**
RC20-174.25A	54.9	−117.6	−15.6	−11.5	0.0
RC20-34-1	57.7	−104.0	−13.8	−11.5	5.9
RC20-34-2	**57.7**	**−97.8**	**−12.9**	**−11.5**	**8.8**
RC20-35	**58.0**	**−65.0**	**−8.4**	**−8.1**	**22.6**
RC20-35a	**58.0**	**−120.5**	**−16.0**	**−15.1**	**−3.7**
RC20-35b	**58.0**	**−82.9**	**−10.9**	**−10.4**	**13.2**
RC20-35avg	58.0	−89.5	−11.8	−11.2	10.7
RC20-93.35	58.1	−114.4	−15.2	−10.9	−0.7
RC20-93.35A	**58.1**	**−73.5**	**−9.6**	**−10.9**	**19.1**
RC20-93.35B	**58.1**	**−116.8**	**−15.5**	**−10.9**	**−1.6**
RC20-40	**59.0**	**−64.9**	**−8.4**	**−11.4**	**27.1**
RC20-40a	59.0	−99.0	−13.1	−11.4	7.9
RC20-40B	**59.0**	**−123.8**	**−16.5**	**−11.4**	**−2.5**
RC20-41a	59.1	−122.8	−16.4	−11.7	−1.3
RC20-41c	59.1	−121.9	−16.2	−11.7	−1.0
RC20-42	**59.3**	**2.1**	**0.7**	**−12.0**	**85.7**
RC20_0.6.7a	**62.2**	**−150.3**	**−20.1**	**−11.7**	**−10.0**
RC20_0-6.7	62.2	−120.5	−16.0	−11.7	−0.5
RC20_0-6.7b	62.2	−105.6	−14.0	−11.7	5.8

Note: Bold items omitted from graphs due to insufficient water or unachievable temperature. *Note:* VSMOW—Vienna standard mean ocean water.

[a] Sample codes begin with the abbreviation for the speleothem and followed by an identification number for the layer number or height from base

[b] Calculated using modern local meteoric water line relationship $\delta D = 7.3\ \delta^{18}O - 3.4$

TABLE 5. RESULTS FROM PALEOTEMPERATURE CALCULATIONS FOR SANTA CRUZ SPELEOTHEM VR-1

Sample ID	Age (ka B.P.)	δD_{fi} (‰) VSMOW	$\delta^{18}O_{fi}$ (‰) (slope 8, d_0=10)[a]	$\delta^{18}O$ (‰) VPDB	T[a]/°C (calculated)	$\delta^{18}O_{fi}$ (‰) (slope 8, d_0=4)[b]	T[b]/°C (calculated)
VR-1-27a	**3.7**	**–75.9**	**–10.7**	**–3.2**	**–9.2**	**–10.0**	**–7.5**
VR-1-27b	**3.7**	**–73.3**	**–10.4**	**–3.4**	**–8.1**	**–9.7**	**–6.3**
VR-1-27	3.7	–44.1	–6.8	–3.3	1.8	–6.0	4.4
VR-1-26-1	4.0	–55.4	–8.2	–3.4	–2.1	–7.4	0.2
VR-1-26-2	4.0	–46.9	–7.1	–3.4	1.0	–6.4	3.6
VR-1-25a	4.3	–36.5	–5.8	–3.3	5.2	–5.1	8.0
VR-1-24	4.6	–29.7	–5.0	–3.3	8.4	–4.2	11.4
VR-1-13	**6.6**	**–68.7**	**–9.8**	**–3.2**	**–7.2**	**–9.1**	**–5.3**
VR-1-13a	6.6	–29.3	–4.9	–3.3	8.6	–4.2	11.6
VR-1-5a	**7.5**	**–51.7**	**–7.7**	**–3.0**	**–2.1**	**–7.0**	**0.2**
VR-1-4	**7.6**	**–54.8**	**–8.1**	**–3.3**	**–2.4**	**–7.4**	**–0.1**
VR-1-3	**7.6**	**–55.4**	**–8.2**	**–3.5**	**–2.0**	**–7.4**	**0.3**
VR-1-2	7.7	–22.7	–4.1	–3.2	11.5	–3.3	14.7

Note: Bold items omitted from graphs due to insufficient water or subzero temperature calculation. VSMOW—Vienna standard mean ocean water; VPDB—Vienna Peedee belemnite.
[a] Calculated using MWL relationship $\delta D = 8\ \delta^{18}O + 10$
[b] Calculated using MWL relationship $\delta D = 8\ \delta^{18}O + 4$

as for the Reed's Cave temperature estimates, the range of realistic temperatures (see above) is 2–12 °C using the modern MWL and 0–11 °C using the modified age MWL. Replicate analyses of sample layer 26 on VR-1 give a temperature difference of 3.4 °C and a δD difference of 8.5‰ (Table 5). This offset is rather large, but is less than the error seen for duplicate analyses of DTAP (standard water) in glass capillaries.

We expected all temperatures from Reed's Cave samples RC2 and RC20 to be lower than the modern value of 10 °C. Temperatures higher than modern or below zero may be due to fractionation as a result of incomplete water recovery. Below zero temperatures may also be due to greater proportion of isotopically depleted water extracted from the smaller (nanometer size) inclusions. Samples with very low yields, usually less than 0.5 µl, also gave negative offsets of up to 40‰ in the analysis of capillary + calcite spar (Table 1). This would lead to lower apparent temperatures of deposition. Samples that yield less than 0.5 µL of water are also not included in the paleotemperature graphs.

As noted above, the $\delta^{18}O_{ct}$ records from the coeval period of growth in samples RC2 and RC20 show an offset in average $\delta^{18}O_{ct}$ and difference in magnitude of isotopic variation (Fig. 7). In a previous paper (Serefiddin et al., 2004) we show how differences in seasonality and flow-paths of recharge can cause such differences. Minor evaporative effects can also cause higher isotopic values in some records with respect to others, but because the deposits were formed in equilibrium (as proved by Hendy test) (Schwarcz, 1986; Hendy, 1971); it is unlikely that evaporative/kinetic fractionation has occurred. Paleotemperatures for these records give similar values and similar direction of change (Fig. 8).

Temperatures were calculated for the two stalagmites from Reed's Cave using the local modern MWL relationship. The coeval part of the record shows a similar direction of change in temperature for both RC2 and RC20. We observe a temperature increase of 7 ± 5 °C in sample RC2 and 8 ± 5 °C in sample RC20 from 62 to 57 ka BP (Table 4). After 57 k.y. B.P. temperatures decrease by 8 ± 5 °C at ca. 54 k.y. B.P. in RC20 and between 57 and 62 k.y. B.P. in RC2. The agreement in magnitude of temperature shift in the two coeval deposits is in striking contrast to the difference in their $\delta^{18}O_{ct}$ records. The magnitude of this temperature shift may, however, be somewhat exaggerated. For comparison, Anderson et al. (2000) observed cooling of up to 10 °C in the Colorado Plateau at the Last Glacial Maximum, suggesting that temperature shifts of this magnitude may have occurred during the last glacial cycle.

The paleotemperature reconstruction for the Santa Cruz speleothem was expected to show a much smaller temperature change

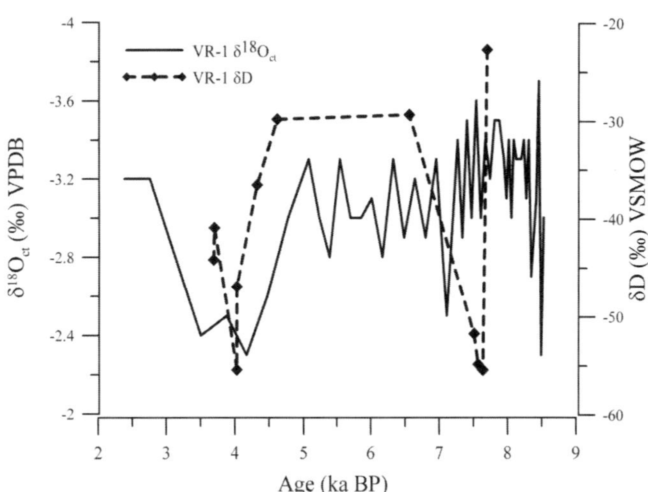

Figure 7. Graph showing δD values of Santa Cruz sample VR-1. VPDB—Vienna Peedee belemnite; VSMOW—Vienna standard mean ocean water.

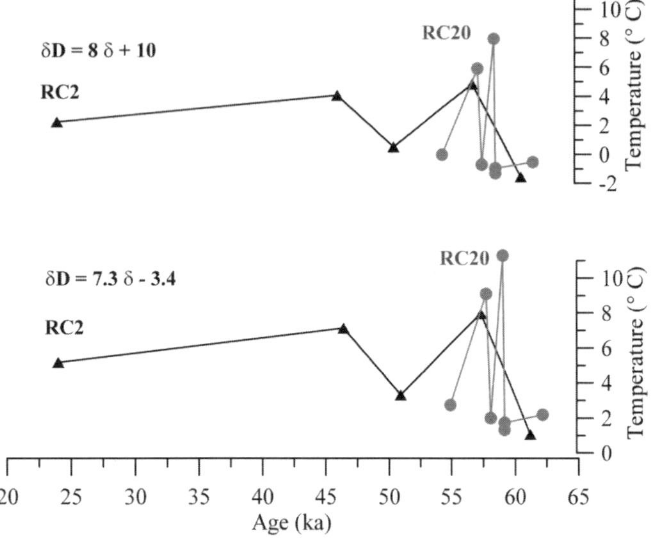

Figure 8. Paleotemperatures for Reed's Cave speleothems 99902 (triangles) and 20000 (circles).

during the Holocene at this climatically moderated coastal location. It is not likely that temperatures would have deviated very much from the modern mean annual temperature of 11 °C. However, temperatures calculated from fluid inclusion-calcite pairs suggest a decrease of almost ~7 ± 2 °C from 4.5 k.y. B.P. to 3.5 k.y. B.P. (Fig. 9). Such a large decrease is not likely, although

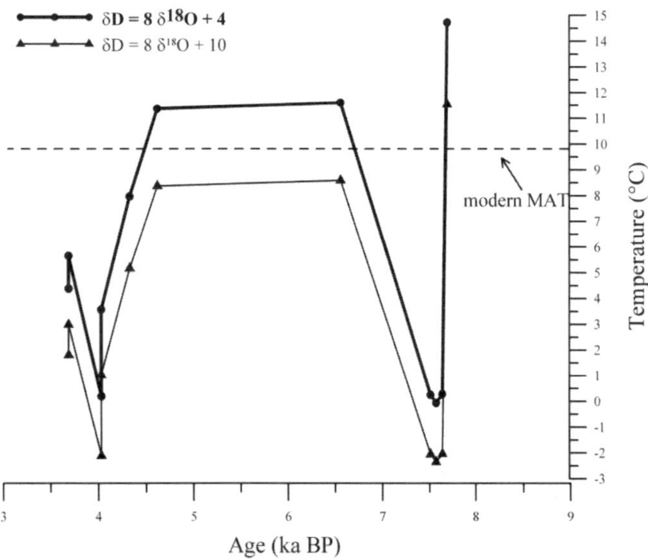

Figure 9. Paleotemperatures for Santa Cruz speleothem VR-1. The bold line uses modified meteoric water line (MWL) relationship (see Table 5) and thin black line uses global MWL relationship. MAT—mean annual temperature.

there is evidence of cooling after 6 ka from marine sediments in the Santa Barbara Basin (Hendy and Kennett, 1999). The warmest temperatures which we obtained were calculated at ca. 8 k.y. B.P., when there is additional evidence for warming from marine sediments (Pisias, 1978). Again, the direction of temperature appears to agree with other proxy records, but the magnitude of change is problematic.

CONCLUSIONS

Although the δD records from Reed's Cave and Santa Cruz appear to be recording some component of global climate change, it is more useful to apply them toward understanding local or regional climate. We have analyzed two coeval stalagmites from Reed's Cave whose $\delta^{18}O_{ct}$ values differed by up to 3‰. If they had been deposited from water of identical isotopic composition, this would have implied a difference of at least 12 °C between their temperatures of deposition, even though they were formed only a few meters apart in the same chamber. The analysis of fluid inclusions from these two deposits confirms there was a corresponding difference in hydrogen isotopic composition of drip waters feeding these deposits, which we assume to have been correlated to corresponding differences in $\delta^{18}O$, and which was the reason for the difference in $\delta^{18}O_{ct}$ between the samples.

It may not be appropriate to use the modern MWL relationship to determine $\delta^{18}O$ values for fluid inclusion samples in the Pleistocene. Assumptions about past MWL relationships must be made to calculate $\delta^{18}O$ values for included waters until the $\delta^{18}O$ of waters can be measured directly. Given the striking difference of the local modern MWL from the global MWL it is however difficult to envision how this relationship may have changed during ice ages. In this paper we have used the modern MWL relationship to demonstrate the agreement in temperature between coeval deposits. This agreement is not very sensitive to the choice of MWLs. We tested this by using the global MWL to calculate $\delta^{18}O$ of the fluid inclusions and again obtained identical temperature shifts around 55 ka for the two stalagmites, and which were of the same magnitude as we obtained using the modern local MWL.

ACKNOWLEDGMENTS

The authors are grateful to the Cave Research Foundation for a grant to F. Serefiddin and the Natural Sciences and Engineering Research Council (Canada) for their support of this research, partly through a Special Project grant "Climate System History and Dynamics" and partly through operating grants to Schwarcz and Ford. Special thanks to P. Rowe and P. Dennis at the University of East Anglia for their assistance with crusher design. George Rossman and Charlie Verdel at the California Institute of Technology provided scanning electron microscope images of nanoinclusions. Greg Stock and Jim Zachos provided sample material from Santa Cruz and unpublished isotope data. Steve Baldwin, Steve Langendorf, Sammi Langendorf, and cavers from the Pahasapa Grotto in the Black Hills assisted with speleothem sample collection and collected all cave water samples.

REFERENCES CITED

Anderson, R.S., 1980, Holocene forest development and paleoclimates within the central Sierra Nevada, California: Journal of Ecology, v. 78, p. 470–489.

Anderson, R.S., Betancourt, J.L., Mead, J.I., Hevly, R.H., and Adam, D.P., 2000, Middle- and late-Wisconsin paleobotanic and paleoclimatic records from the southern Colorado Plateau, USA: Palaeogeography, Palaeoclimatology, Palaeoecology, v. 155, p. 31–57, doi: 10.1016/S0031-0182(99)00093-0.

Bar-Matthews, M., Gilmour, M., Ayalon, A., Vax, A., Frumkin, A., and Hawkesworth, C., 2000, Variation of palaeoclimate in the Eastern Mediterranean Region—As derived from speleothems in various climate regimes in Israel: Goldschmidt 2000, Journal of Conference Abstracts, v. 5, no. 2, p. 194–195.

Craig, H., 1961, Isotopic variations in meteoric waters: Science, v. 133, p. 1702.

Dansgaard, W., 1964, Stable isotopes in precipitation: Tellus, v. 16, p. 438–468.

Dennis, P.F., Rowe, P.J., and Atkinson, T.C., 2001, The recovery and isotopic measurement of water from fluid inclusions in speleothems: Geochimica et Cosmochimica Acta, v. 65, no. 6, p. 871–884, doi: 10.1016/S0016-7037(00)00576-7.

Dorale, J.A., 2000, A high-resolution record of climate and vegetation change from Crevice Cave, Missouri during the last interglacial-glacial cycle [Ph. D. thesis]: Minneapolis, University of Minnesota, 300 p.

Dorale, J.A., Edwards, R.L., Ito, E., and González, L.A., 1998, Climate and vegetation history of the Midcontinent from 75 to 25 ka: A Speleothem Record from Crevice Cave, Missouri, USA: Science, v. 282, p. 1871–1874, doi: 10.1126/science.282.5395.1871.

Driscoll, D.G., Carter, J.M., Williamson, J.E., and Putnam, L.D., 2002, Hydrology of the Black Hills Area, South Dakota: U.S. Geological Survey Water Resources Investigation Report 02–4094.

Gagan, M.K., Ayliffe, L.K., Hopley, D., Cali, J.A., Mortimer, G.E., Chappell, J., McCulloch, M.T., and Head, M.J., 1998, Temperature and surface-ocean water balance of the mid-Holocene tropical Western Pacific: Science, v. 279, p. 1014–1018, doi: 10.1126/science.279.5353.1014.

Genty, D., Plagnes, V., Causse, C., Cattani, O., Stievenard, M., Falourd, S., Blamarr, D., Ouahdi, R., and Van-Exter, S., 2002, Fossil water in large stalagmite voids as a tool for paleoprecipitation stable isotope composition reconstitution and paleotemperature calculation: Chemical Geology, v. 184, p. 83–95, doi: 10.1016/S0009-2541(01)00356-4.

Goede, A., Green, D.C., and Harmon, R.S., 1986, Late Pleistocene palaeotemperature record from a Tasmanian speleothem: Australian Journal of Earth Sciences, v. 33, p. 333–342.

Goede, A., Veeh, H.H., and Ayliffe, L.K., 1990, Late Quaternary paleotemperature records for two Tasmanian speleothems, Australian Journal of Earth Sciences, v. 37, p. 267–278.

Harmon, R.S., and Schwarcz, H.P., 1981, Changes in ^{2}H and ^{18}O enrichment of meteoric water and Pleistocene glaciation: Nature, v. 290, p. 125–128, doi: 10.1038/290125a0.

Harmon, R.S., Schwarcz, H.P., and O'Neil, J.R., 1979, D/H ratios in speleothem fluid inclusions: a guide to variations in the isotopic composition of meteoric precipitation?: Earth and Planetary Science Letters, v. 42, p. 254–266, doi: 10.1016/0012-821X(79)90033-5.

Hendy, C., 1971, The isotopic geochemistry of speleothems I. The calculation of the effects of different modes of formation on the isotopic composition of speleothems and their applicability as palaeoclimatic indicators: Geochimica et Cosmochimica Acta, v. 35, p. 801–824, doi: 10.1016/0016-7037(71)90127-X.

Hendy, I.L., and Kennett, J.P., 1999, Latest Quaternary North Pacific surface-water responses imply atmosphere-driven climate stability: Geology, v. 27, p. 291–294, doi: 10.1130/0091-7613(1999)027<0291:LQNPSW>2.3.CO;2.

International Atomic Energy Agency, 2001, GNIP Maps and Animations: International Atomic Energy Agency, Vienna (accessible at http://isohis.iaea.org).

Jouzel, J., Lorius, C., Petit, J.R., Genthon, C., Barkov, N.I., Kotlyakov, V.M., and Petrov, V.M., 1987, Vostok ice core: a continuous isotope temperature record over the last climatic cycle (160,000 years), Nature, v. 329, p. 403–407, doi: 10.1038/329403a0.

Longinelli, A., and Selmo, E., 2003, Isotopic composition of precipitation in Italy: a first overall map: Journal of Hydrology, v. 270, no. 1-2, p. 75–88, doi: 10.1016/S0022-1694(02)00281-0.

Martinson, D.G., Pisias, N.G., Hays, J.D., Imbrie, J., Moore, T.C., Jr., and Shackleton, N.J., 1987, Age dating and the orbital theory of the Ice Ages: development of a high-resolution 0–300,000 year chronostratigraphy: Quaternary Research, v. 27, p. 1–29, doi: 10.1016/0033-5894(87)90046-9.

Matthews, A., Ayalon, A., and Bar-Matthews, M., 2000, D/H ratios of fluid inclusions of Soreq Cave (Israel) speleothems as a guide to the Eastern Mediterranean Meteoric Water Line relationships in the last 120 Ky: Chemical Geology, v. 166, p. 183–191, doi: 10.1016/S0009-2541(99)00192-8.

McDermott, F., Fairchild, I., Schwarcz, H.P., and Rowe, P.J., 2005, Isotopes in speleothems, *in* Leng, M.J., ed., Isotopes in Paleoenvironmental Research: Developments in Paleoenvironmental Research, Volume 10: Kluwer (in press).

McGarry, S., Bar-Matthews, M., Matthews, A., and Ayalon, A., 2002, Palaeohydrology in the Eastern Mediterranean from speleothem fluid inclusion D/H analyses: Geochimica et Cosmochimica Acta, v. 66, no. 1, p. A500.

Merlivat, M., and Jouzel, J., 1969, Global climatic interpretation of the deuterium–oxygen-18 relationship for precipitation: Journal of Geophysical Research, v. 84, p. 5029–5033.

Nativ, R., and Riggio, R., 1990, Precipitation in the Southern High Plains: Meteorologic and isotopic features: Journal of Geophysical Research, v. 95, no. D13, p. 22559–22564.

O'Neil, J.R., Clayton, R.N., and Mayeda, T.K., 1969, Oxygen isotope fractionation in divalent metal carbonates: The Journal of Chemical Physics, v. 51, p. 5547–5558, doi: 10.1063/1.1671982.

Pisias, N., 1978, Paleoceanography of the Santa Barbara Basin during the last 8000 years: Quaternary Research, v. 10, p. 366–384.

Rozanski, K., and Dulinski, M., 1987, Deuterium content of European paleowaters as inferred from isotopic composition of fluid inclusions trapped in carbonate cave deposits: IAEA-SM-299/99, Vienna, p. 565–578.

Rozanski, K., Araguás-Araguás, L., and Gonfiantini, R., 1992, Relation between long-term trends of oxygen-18 isotope composition of precipitation and climate: Science, v. 258, p. 981–985.

Rozanski, K., Araguás-Araguás, L., and Gonfiantini, R., 1993, Isotopic patterns in modern global precipitation, *in* Swart, P.K., Lohmann, K.C., McKenzie, J., and Savin, S., eds., Climate Change in Continental Isotopic Records: American Geophysical Union Geophysical Monograph 78, p. 1–36.

Schwarcz, H.P., 1986, Geochronology and isotope geochemistry in speleothems, *in* Fritz, P., and Fontes, J., eds., Handbook of Environmental Isotope Geochemistry: Amsterdam, Elsevier Publishers, p. 271–303.

Schwarcz, H.P., and Yonge, C.J., 1983, Isotopic composition of paleowaters as inferred from speleothem and its fluid inclusions, *in* International Atomic Energy Agency, Paleoclimates and Paleowaters: A collection of Environmental Isotope studies, IAEA STI/PUB/621, p. 115–133.

Schwarcz, H.P., Harmon, R.S., Thompson, P., and Ford, D.C., 1976, Stable isotope studies of fluid inclusions in speleothems and their paleoclimatic significance: Geochimica et Cosmochimica Acta, v. 40, p. 637–665.

Serefiddin, F., Schwarcz, H.P., Ford, D.C., and Baldwin, S., 2004, Late Pleistocene paleoclimate in the Black Hills of South Dakota from isotope records in speleothems: Palaeogeography, Palaeoclimatology, Palaeoecology, v. 203, p. 1–17, doi: 10.1016/S0031-0182(03)00639-4.

Thompson, R.S., Whitlock, C., Bartlein, P.J., Harrison, S.P., and Spaulding, W.G., 1993, Climatic changes in the Western United States since 18,000 yr BP, *in* Wright, H.P., et al., eds., Global climates since the Last Glacial Maximum: University of Minnesota Press, p. 468–514.

Yonge, C.J., 1982, Stable isotope studies of water extracted from speleothems [Ph.D. thesis]: Hamilton, Ontario, Canada, McMaster University, 144 p.

Yonge, C.J., Ford, D.C., Gray, J., and Schwarcz, H.P., 1985, Stable isotope studies of cave seepage water: Chemical Geology, v. 58, p. 97–105, doi: 10.1016/0009-2541(85)90182-2.

Manuscript Accepted by the Society 19 April 2005

Printed in the USA

Geological Society of America
Special Paper 395
2005

Juxtaposed Permian and Pleistocene isotopic archives: Surficial environments recorded in calcite and goethite from the Wichita Mountains, Oklahoma

Neil J. Tabor
Crayton J. Yapp
Department of Geological Sciences, Southern Methodist University, Dallas, Texas 75725-0395, USA

ABSTRACT

A paleokarst fill deposit from the Wichita Mountains, south-central Oklahoma, United States, consists primarily of sparry calcite, Fe-sulfides, and goethite. Previous cement-stratigraphic studies and paleontological finds suggest that calcite mineralization was initiated during Permian time, whereas goethite and other oxides apparently formed from oxidation of preexisting Fe-sulfides during Pleistocene time. Therefore, these deposits have the potential to offer insight into surficial hydrology and paleoenvironment in an upland setting from two time periods at a single site.

$\delta^{13}C_{PDB}$ and $\delta^{18}O_{SMOW}$ measurements of 17 samples from growth bands in a single karst-fill calcite crystal range from –10.7‰ to –6.6‰ (mean = –8.6‰) and 27.1‰ to 28.3‰ (mean = 28‰), respectively. Large oscillations in the $\delta^{13}C$ values through the growth series may originate from seasonal changes in the magnitude of biological productivity during Permian time. These $\delta^{13}C$ oscillations contrast with the relative stability of the $\delta^{18}O$ values, which are more positive than would be expected for isotopic equilibrium with local modern waters. The $\delta^{18}O$ values of the calcite may reflect the $\delta^{18}O$ values of ambient meteoric groundwaters in the Permian that were isotopically similar to waters in modern, seasonally dry, low-latitude coastal regions.

Goethites are not in equilibrium with modern waters or coexisting calcites in the fissure-fill deposit as determined from $\delta^{18}O$ and δD values of the goethites. Furthermore, the combined $\delta^{18}O$ and δD values of the goethites are indicative of formation from meteoric waters at a temperature of ~9 °C ± 3 °C. This inferred temperature is 7 °C ± 3 °C cooler than local modern mean annual temperature and corresponds well with independent studies that propose temperatures ~6 °C cooler in this region during Pleistocene time.

The mole fraction and $\delta^{13}C$ values of the $Fe(CO_3)OH$ component in solid solution in the goethite sample are 0.0103 and –10.1‰, respectively. In combination, these values suggest that goethite formed in an environment characterized by mixing of three isotopically distinct CO_2 components: (1) oxidized biological carbon, (2) atmospheric CO_2, and (3) CO_2 from dissolution of carbonate in the karst system. Oxidized biological carbon may have originated either from flora characterized by C3 or mixed C3:C4 photosynthesizers. Mass balance calculations between these three CO_2 end members correspond to an inferred soil CO_2 concentration [CO_2 contributed from (1) and (2) above] ranging from ~8,000 ppmV to ~16,000 ppmV for a local ecosystem dominated by C_3 flora. This inferred range of soil CO_2 concentrations is typical of grasslands characterized by relatively high biological productivity. If C4 flora were a significant

Tabor, N.J., and Yapp, C.J., 2005, Juxtaposed Permian and Pleistocene isotopic archives: Surficial environments recorded in calcite and goethite from the Wichita Mountains, Oklahoma, *in* Mora, G., and Surge, D., eds., Isotopic and elemental tracers of Cenozoic climate change: Geological Society of America Special Paper 395, p. 55–70, doi: 10.1130/2005.2395(06).

source of oxidizing carbon, the higher calculated ambient CO_2 concentration at the time of goethite crystallization in the cave (~20,000 ppmV) might be interpreted to correspond to an unusually productive C4 soil present at a time of generally cooler and drier conditions across the southern Great Plains of North America.

Keywords: Permian; Pleistocene; P_{CO_2}; oxygen, hydrogen and carbon isotopes, paleoclimate.

INTRODUCTION

The relationship between oxygen and hydrogen isotopes in meteoric waters [the Meteoric Water Line (MWL) of Craig, 1961] has served as a basis for using single-mineral geothermometers in hydroxyl minerals to deduce paleotemperatures (Savin and Epstein, 1970; Yapp, 1987, 1993, 2000; Delgado and Reyes, 1996; Tabor et al., 2004a). Furthermore, isotopic analyses of secondary alteration minerals preserved in nearly continuous stratigraphic successions have been used to infer regional Cretaceous and Late Paleozoic paleoatmospheric circulation patterns (Ludvigson, 1998; Tabor and Montañez, 2002) and to monitor the long-term effects (10^6 to 10^8 m.y.) of tectonic uplift (Poage and Chamberlain, 2001) and continental drift (e.g., Bird and Chivas, 1988; Lawrence and Rashkes-Meaux, 1993; Gilg, 2000). In the current study, we present data from juxtaposed Permian and Pleistocene minerals that contain isotopic records of contrasting surficial environments.

The area is centered around the Wichita Mountains of south-central Oklahoma and consists of karsted Ordovician limestones that have been upland (or piedmont) terrain since early Permian time (Donovan et al., 2001). The Ordovician karst hosts two temporally distinct groups of secondary minerals: (1) early Permian calcite and Fe-sulfide and (2) Pleistocene Fe (III) oxides (hydroxides) formed by oxidative dissolution of preexisting sulfides. Hydrogen-, oxygen-, and carbon-isotope analyses of calcites and goethites from a cave-fill deposit in the Wichita Mountains, south-central Oklahoma, United States, are discussed in terms of their implications for ambient Pleistocene and Permian paleoenvironments.

BACKGROUND

The Wichita Mountains region of southern Oklahoma appears to be part of a failed ancient continental rift (aulacogen; Donovan, 1986). Initial heat flux and lithospheric expansion in this region with attendant intrusive igneous activity probably began in late Precambrian–Cambrian time, followed by lithospheric cooling, basin subsidence, and deposition of thousands of meters of carbonate and siliciclastic strata during the Late Cambrian, Ordovician, and Carboniferous (Donovan, 1986). A transpressive-stress regime, related to the assembly of Pangea in the late Carboniferous and early Permian, created several asymmetric basins and uplifts before the aulacogen stabilized and tectonic quiescence ensued in middle Permian time. These uplifts include the Wichita and Arbuckle Mountain chains.

Displacement along the Meers and Mountain View faults, bounding the northern edge of the eastern Wichita Mountains, exhumed and exposed Ordovician-age marine carbonate strata to surficial terrestrial weathering during Late Pennsylvanian and early Permian time. Physical weathering of the Ordovician deposits resulted in fluvial deposition of the Post-Oak Conglomerate, whereas chemical weathering resulted in widespread karstification, as indicated by numerous and small, back-filled cave systems (Donovan et al., 2001). Both the Post-Oak Conglomerate and cave systems are found in the Slick Hills region of southwestern Oklahoma (Fig. 1).

Back-filling of the cave systems in this karsted terrain owes its origins to surface-tied hydrology indicative of a shallow burial history (Donovan et al., 2001). In addition, late Paleozoic vertebrate fossils preserved within authigenic sulfide-bearing calcareous rocks in the fissure-fill deposits indicate a Permian (Leonardian to ~Artinskian) age for these fissure-fill deposits (Olson, 1967). Subsequent erosion of a thin sedimentary overburden in Pleistocene time led to oxidation of Fe-sulfides, dissolution of carbonate, and crystallization of secondary Fe (III) oxides and oxyhydroxides (Donovan et al., 2001; 2003, personal commun.).

METHODS

The karst-fill deposit in this study is exposed along Highway 58, in Blue-Creek Canyon, in Ordovician-age marine limestones of the upper Arbuckle Group. The karst deposit is vertically oriented, with a height of ~6 m and a width of as much as 1 m (Fig. 2A). Internally, the fissure-fill deposit is characterized by two lithologies: (1) a 5–25-cm-thick red-to-brown iron-oxide–rich layer that lines the contact with the karsted Ordovician limestone and (2) sparry calcite veins and "pockets" that range from 10 cm to 35 cm thick (Fig. 2B).

Fissure-Fill Calcite and Ordovician Limestone

A sparry calcite crystal from the fissure fill and a sample of the Ordovician limestone that hosts the fissure fill deposits were sectioned and polished (Fig. 3). Samples from a number of different "growth bands" in the fissure-fill calcite were taken directly from the polished slab using a cleaned hand-held metal probe. The Ordovician limestone was sampled using a hand-held dental drill with faceted burr and 100 μm bit. Approximately 50 μg of carbonate powder were roasted at 375 °C in vacuum for three hours to remove organics. Carbonate samples were then reacted

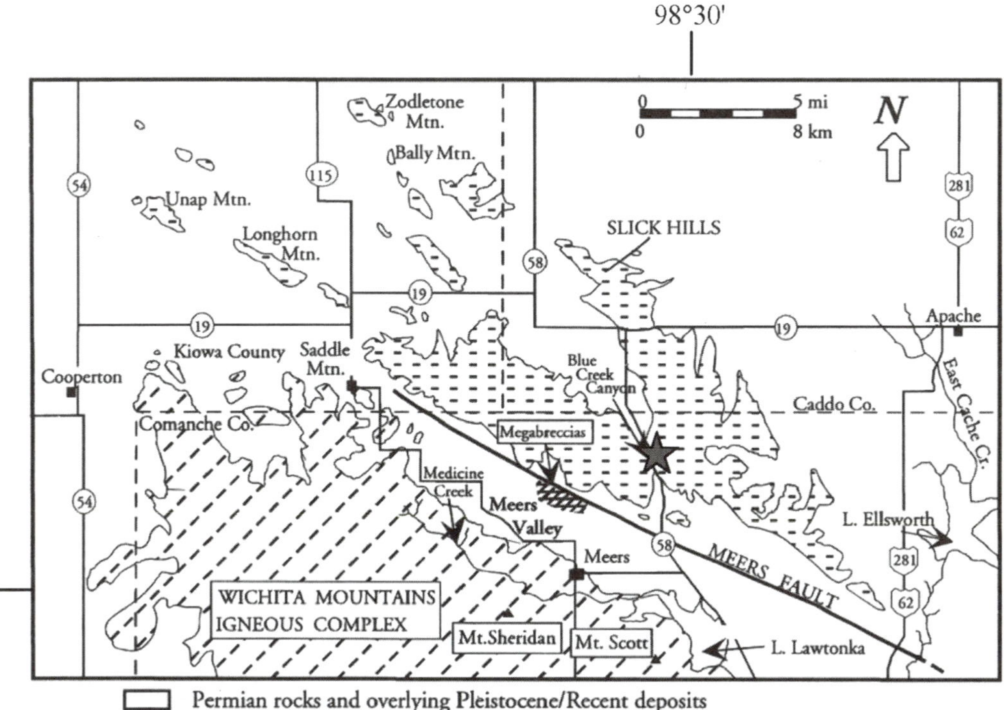

Figure 1. Regional map of south-central Oklahoma (after Donovan et al., 2001). The area denoted by the arrow is the general vicinity of the cave-fill deposit that is the focus of this study.

Figure 2. (A) Outcrop photograph of the cave-fill deposit. Dan Chaney (~155 cm tall) for scale. (B) Close-up photograph showing the Fe-oxide (orange) and calcite dominated (white) areas of the cave-fill deposit. Hammer for scale. See text for discussion of features.

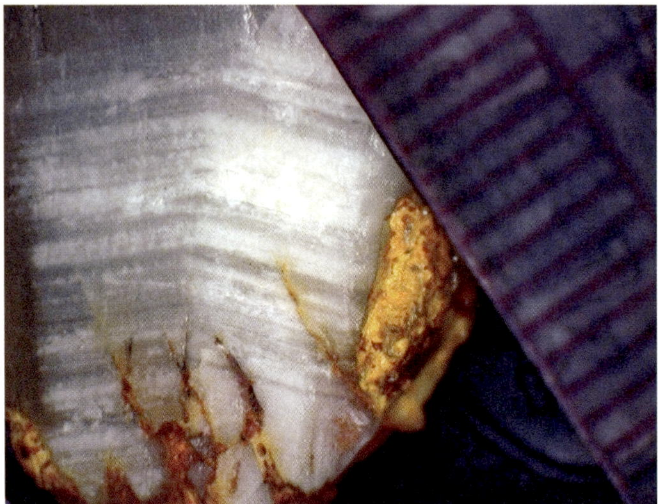

Figure 3. Reflected-light photomicrograph of a sparry calcite sample from the cave-fill deposit. Light-colored bands have a high concentration of fluid inclusions, whereas darker-colored bands do not contain fluid inclusions. Note the occurrence of Fe-oxide coatings on the exterior of the sample and dissolution pitting near the base of the photograph. Demarcations on ruler are 1 mm apart. See text for discussion.

at 90 °C with 100% H_3PO_4 to produce CO_2. $\delta^{13}C$ and $\delta^{18}O$ analyses of CO_2 evolved from all calcite samples were carried out on a Fisons-Optima IR gas source mass spectrometer in the Department of Geology at the University of California, Davis.

In this work, $\delta^{13}C$, $\delta^{18}O$, and δD values of CO_2 and H_2 derived from carbonates and goethites are given relative to the Peedee belemnite (PDB) standard for carbon (Craig 1957) and the Vienna standard mean ocean water (V-SMOW) standard for hydrogen and oxygen isotopes (Gonfiantini 1984), where

$$\delta^{13}C, \delta^{18}O, \delta D = (R_{sample}/R_{standard} - 1) \times 1000‰ \quad (1)$$

and R = $^{13}C/^{12}C$, $^{18}O/^{16}O$, D/H, respectively. Based on 19 measurements of NBS-19 over the period of analysis, $\delta^{13}C$ and $\delta^{18}O$ measurements of CO_2 evolved from calcite are ±0.1‰ and ±0.2‰, respectively.

Iron Oxides

After collection from the karst-fill deposit, two iron-oxide–rich samples, labeled OKCave and OKCaveM, were transported to the laboratory wrapped in aluminum foil. The iron-oxide samples were prepared for isotopic analysis using the methods of Yapp (1998). Samples were ground in an aluminum oxide mortar and pestle under reagent-grade acetone and sized by passage through a 63 µm brass sieve. Only powders from the <63 µm particle size fraction were used in this study. Powdered samples were treated overnight with ~40 mL of 0.5N HCl solution to remove admixed carbonates and then rinsed with successive aliquots of deionized H_2O until the pH of the rinse water was equivalent to the initial pH of the deionized H_2O. Each sample was subsequently treated over a period of 28–35 days with successive 40 mL aliquots of 30% H_2O_2 at room temperature in order to promote oxidation of any admixed organic matter that may have been present within the samples. As the reactivity of the solution diminished and the suspended particles settled, the solution was decanted and replaced by a fresh aliquot of H_2O_2. After 12–14 H_2O_2 treatments, samples were dried in a vacuum desiccator at room temperature. In this paper, samples subjected only to the foregoing treatments are designated bulk samples.

Mineralogy was determined by X-ray diffraction (XRD) analysis of powdered samples using Cu-Kα radiation on a Diano 8500 X-ray diffractometer in the Department of Land, Air and Water Resources at the University of California, Davis. Powders were backmounted into an aluminum holder and step-scanned from 2 to 70° 2θ with 0.01° steps, a dwell time of 12 s, 40 kV, and 20 mA, 1° divergence and scatter slits, and a 0.2 mm receiving slit. The amount of Al substituted for Fe in goethite was determined by the XRD method of Schulze (1984). This XRD–based calculation for Al^{3+} substitution has an analytical uncertainty of ±3 mol%.

Chemical Analysis

For chemical analyses, bulk samples of OKCave and OKCaveM were combined with lithium tetraborate to produce a 2:1 mixture on a mass basis. These mixtures were fused in graphite crucibles at temperatures of 1000 °C for one hour and then quenched in deionized water to produce a glass that was subsequently ground to <63 µm. For each sample, ~125 mg of fused glass was sealed in 15 mL Teflon bombs with 10 mL of concentrated HNO_3 and left on a hot plate at 100 °C until all of the solids dissolved. Dissolution of the samples occurred within 2 days. Each 10 mL aliquot was then transferred to 100 mL volumetric flasks, and the solution was diluted to 2%–3% HNO_3. Chemical analyses of the dilute solutions of HNO_3^+ sample were performed on an inductively coupled plasma–optical emission spectroscopy (ICP-OES) at the DANR Analytical Facility at the University of California, Davis. The relative analytical error of these analyses is no greater than ±2% of the reported value for the oxide component.

In order to assess the isotopic composition of non–iron oxide constituents within the samples, ~200 mg aliquots of bulk samples were subjected to citrate-bicarbonate-dithionite (CBD) digestion solutions to remove iron oxides (Jackson, 1979). The remaining residue was then washed with five to eight successive 50 mL aliquots of deionized H_2O and subsequently treated with three to four successive 40 mL aliquots of 30% H_2O_2 in order to facilitate oxidation of any additional organics that may have become accessible as a result of CBD treatment. After H_2O_2 treatment, the residues were dried at room temperature in a vacuum desiccator. In this paper, these non–iron oxide constituents are designated "residue" samples.

Structural oxygen was extracted from the bulk and residue samples and quantitatively converted to CO_2 with BrF_5 in the

TABLE 1A. DATA FROM INCREMENTAL DEHYDRATION OF OKCAVEMICRO

Time (min)	T (°C)	H_2 yield (μmol)	CO_2 Yield (μmol)	CO_2 $\delta^{13}C$	$X_v(H_2)$	F
Initial goethite mass = 305.0mg						
30*	210	468	5.2	−12.9	0.27	0.0111
30	210	746	7.3	−10.1	0.69	0.0098
30	210	200	2.4	−10.0	0.80	0.0120
30	210	63	1.0	−11.6	0.84	0.0159
108	210	41	0.5	−22.1	0.86	0.0122
30*	850	247	10.2	−16.9	1.00	0.0413

*Closed system dehydration in ~0.16 bar of O_2. All other increments open system in vacuum.

TABLE 1B. PLATEAU VALUES FOR OKCAVEMICRO

F	X_m*	$1/X_m$	$\delta^{13}C$
0.0103	0.0052	192	−10.1‰

*X = 0.5F for plateau values and X = the mol fraction of $Fe(CO_3)OH$.

Department of Geological Sciences at the Southern Methodist University using the procedure of Clayton and Mayeda (1963). The oxygen-isotope composition of the resultant CO_2 was measured on a Finnigan MAT 252 isotope ratio mass spectrometer in the Department of Geological Sciences at the Southern Methodist University.

For D/H (dueterium/hydrogen; H^2/H^1) analysis, samples were initially outgassed under vacuum at ~120 °C for ~10 h to remove adsorbed water. Mineral-bound hydrogen for δD analysis was extracted as H_2O by heating the samples to ~1100 °C under vacuum in closed-system conditions. The liberated structural water was then converted to H_2 by passage over hot (~750 °C) U-metal (Bigeleisen, 1952) at the Southern Methodist University. The hydrogen yield was determined manometrically, and the δD values of evolved gases were measured with a Finnigan MAT 252 at the Southern Methodist University. Based on the range of values for a single sample, the reported δD values have an analytical uncertainty of about ±3‰.

Ferric Carbonate Analysis

Abundance and $\delta^{13}C$ values for the iron (III) carbonate ($Fe(CO_3)OH$) component in solid solution in goethite were measured at the Department of Geological Sciences at the Southern Methodist University on samples treated with 0.5N HCl, followed by deionized H_2O, then 30% H_2O_2, using the published incremental dehydration-decarbonation methods of Yapp and Poths (1991, 1993). The detailed results for each incremental dehydration-decarbonation procedure are given in Table 1. The CO_2 and H_2O collected at each step were separated cryogenically. The evolved water was quantitatively converted to H_2 over U-metal at ~750 °C. Yields of H_2 were measured manometrically with a precision of about ±1 μmol. For CO_2, differences in yield of ~0.1 μmol could be resolved. Amounts of CO_2 as low as 0.1 μmol can be evolved from the $Fe(CO_3)OH$ in a single step during incremental dehydration-decarbonation. Samples of evolved CO_2 were analyzed for $\delta^{13}C$ values on a Finnigan MAT 252 mass spectrometer at the Southern Methodist University. The analytical uncertainty of the $\delta^{13}C$ value of $Fe(CO_3)OH$ is about ±0.2‰ (Yapp and Poths, 1993).

RESULTS

Calcite-Rich Fissure Fill

Calcite-spar crystals from the fissure-fill deposit exhibit drusy crystal morphologies (Fig. 3) with ~1-mm-thick laminations of alternating clear and cloudy bands oblique to the crystallographic c-axis. The layers with a cloudy appearance are a result of relatively high concentrations of fluid inclusions. The exterior of this sample, which was taken from the interface between the oxide-rich and calcite-rich parts of the fissure-fill deposit, exhibits dissolution pitting and microkarst fabrics (Fig. 3). In addition, a number of flowstone and stalactite samples from the fissure-fill deposits also locally contain laminations with abundant sulfide minerals intercalated with calcite (Fig. 4).

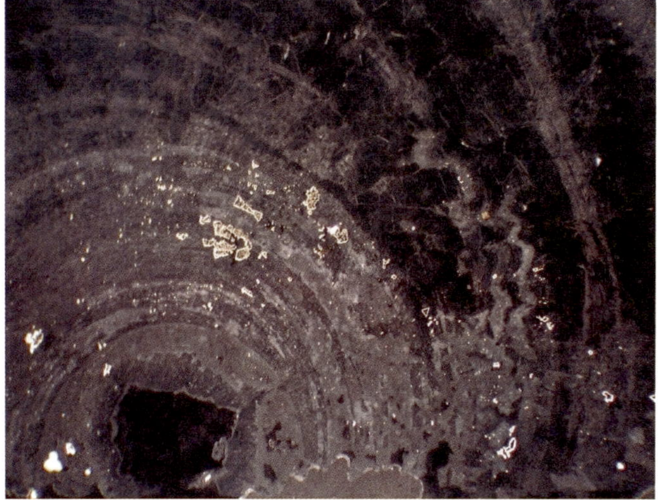

Figure 4. Reflected light photomicrograph of a speleothem from a cave-fill deposit. The speleothem is primarily composed of calcite with subordinate Fe-sulfides (white areas). The long field of view in this figure is 1cm. See text for discussion.

The measured carbon- and oxygen-isotope compositions from a series of cloudy and clear growth bands in the calcite-spar crystal (Fig. 3) are presented in Table 2 and Figure 5. Oxygen-isotope compositions exhibit fairly uniform and positive $\delta^{18}O_{SMOW}$ values ranging from 27.1‰ to 28.3‰ ($\delta^{18}O_{PDB}$ = –3.7‰ to –2.5‰, n = 17), with a mean isotopic composition of 28.0‰ ($\delta^{18}O_{PDB}$ = –2.8‰ ± 0.3‰, 1s) (Table 2 and Fig. 5). $\delta^{13}C_{PDB}$ values are more variable, ranging from –6.6‰ to –10.7‰, with a mean isotopic composition of –8.6‰ ± 1.5‰ (1s, n = 17). The $\delta^{13}C_{PDB}$ and $\delta^{18}O_{PDB}$ values of Ordovician marine limestone associated with the karst are 1.5 ± 0.1‰ and –1.2 ± 0.2‰ (n = 2), respectively.

TABLE 2. CARBON AND OXYGEN ISOTOPE COMPOSITIONS FOR A CALCITE-SPAR CRYSTAL FROM THE FISSURE FILL DEPOSIT

Growth Band #	$\delta^{13}C_{PDB}$	$\delta^{18}O_{PDB}$	$\delta^{18}O_{SMOW}$
1	–6.6	–3.7	27.1
2	–10.1	–2.7	28.1
3	–10.1	–2.8	28.0
4	–9.1	–2.9	27.9
5	–7.4	–2.7	28.1
6	–6.9	–2.6	28.2
7	–9.0	–3.1	27.7
8	–9.3	–2.7	28.1
9	–9.4	–2.8	28.0
10	–10.7	–2.6	28.2
11	–8.9	–2.7	28.1
12	–7.5	–2.5	28.3
13	–6.7	–2.9	27.9
14	–10.2	–2.8	28.0
15	–6	–2.7	28.1
21	–10.0	–2.8	28.0
41	–9.1	–2.9	27.9
Avg.	–8.6	–2.8	28.0
St. dev. (1σ)	1.5	0.3	0.3

Iron Oxide–Rich Fissure Fill

XRD analyses of the OKCave and OKCaveM samples are essentially identical, with high and broad peaks at low angle reflections of 5.96° and 12.52° 2θ (Fig. 6). These peaks likely correspond to the basal (001) spacings of 2:1 and 1:1 phyllosilicates, respectively (Moore and Reynolds, 1997). High angle reflections at 17.88°, 21.28°, 33.28°, and 36.72° 2Θ correspond to crystallographic d(*hkl*) spacings of goethite (α-FeOOH). Goethite (111) and (110) d-spacings of 2.448Å and 4.183Å for both OKCaveM and OKCave indicate 3 ± 3 mol% Al^{3+} substitution for Fe^{3+} in goethite in both samples (Schulze, 1984).

Results of the chemical analyses for OKCaveM and OKCave are reported in Table 3 as the mole fraction of the oxide components. $X(O)_{Fe}$ values are the calculated amount of oxygen in stoichiometric goethite as a mole fraction of the total oxygen in the bulk sample. Also reported in Table 3 is the value $X(O)_{Al}$, which is the amount of oxygen in the goethite crystal lattice associated with Al^{3+} substitution as a mole fraction of the total oxygen in the goethite. The value of $X(O)_{Al}$ was determined from X-ray diffraction analysis (Schulze, 1984).

Measured $\delta^{18}O$ and δD values for the bulk and residue samples are presented in Table 4. In both cases, the residues remaining after complete or partial dissolution of goethite and hematite have more positive $\delta^{18}O$ and δD values than the bulk (Fe-oxide–rich) fractions. The $X(O)_{Fe}$ and $X(O)_{Al}$ values reported in Table 3 were used to calculate the end-member oxygen-isotope compositions of goethite in OKCave and OKCaveM for both samples following the mass-balance approach of Yapp (1998), in which

$$\delta^{18}O_{bulk} = X(O)_{Fe}*\delta^{18}O_{Fe} + X(O)_{Al}*\delta^{18}O_{Al} + X(O)_{residue}*\delta^{18}O_{residue} \quad (2a)$$

$$1 = X(O)_{Fe} + X(O)_{Al} + X(O)_{residue} \quad (2b)$$

$\delta^{18}O_{residue}$ was measured directly, whereas the value for $X(O)_{residue}$ was determined from the chemical data presented in Table 3 as the remaining fraction of oxygen after subtraction of $X(O)_{Fe}$ and $X(O)_{Al}$ contributed from goethite to the total oxygen.

The hydrogen-isotope compositions of end-member goethite samples were determined by a mass balance calculation similar to that of the oxygen calculations. The D/H calculation considers the differences between the measured wt% H_2O and the δD values of the bulk and residue samples as well as the calculated mole fraction of hydrogen $X(H)_{Fe}$ in goethite. These end-member oxygen- and hydrogen-isotope compositions of goethite are reported in Table 4.

The results of the incremental vacuum dehydration-decarbonation analyses of OKCaveM goethite are given in Table 1 and depicted as incremental dehydration-decarbonation spectra in Figure 7. The progress variable $X_v(H_2)$ is the cumulative sum of evolved H_2 as a mole fraction of the total hydrogen in the goethite sample. When $X_v(H_2) = 0$, there has been no breakdown of goethite. When $X_v(H_2) = 1$, the goethite has been completely converted to hematite (Yapp and Poths, 1993). The "F" parameter in Figure 7A is defined as follows (Yapp and Poths, 1993): $F = n(CO_2)/n(H_2O)$, where $n(CO_2)$ = μmol of CO_2 evolved in an increment of goethite dehydration-decarbonation and $n(H_2O)$ = μmol of H_2O evolved over that same increment.

Previous work has shown that, for $X_v(H_2)$ values from ~0.2 or 0.3 to ~0.6 or 0.8, values of F and $\delta^{13}C$ commonly exhibit "plateaus" for which there is little change in either $\delta^{13}C$ or F as a function of $X_v(H_2)$ (Yapp and Poths, 1991, 1993; Yapp, 1997; Hsieh and Yapp, 1999). Such plateaus are produced by the CO_2 evolved from $Fe(CO_3)OH$ in the goethite structure, because $Fe(CO_3)OH$ only breaks down when the local, confining goethite structure is destroyed (Yapp and Poths, 1991). The 850 °C step (and, at times, the longer duration vacuum dehydration steps that immediately precede the 850 °C step) commonly does not exhibit these plateau values, because this increment may include CO_2 from oxidation of refractory organic matter (Yapp and Poths, 1991, 1992, 1993). Average plateau $\delta^{13}C$ and F values represent the weighted mean of CO_2 and H_2O incremental values over the plateau interval

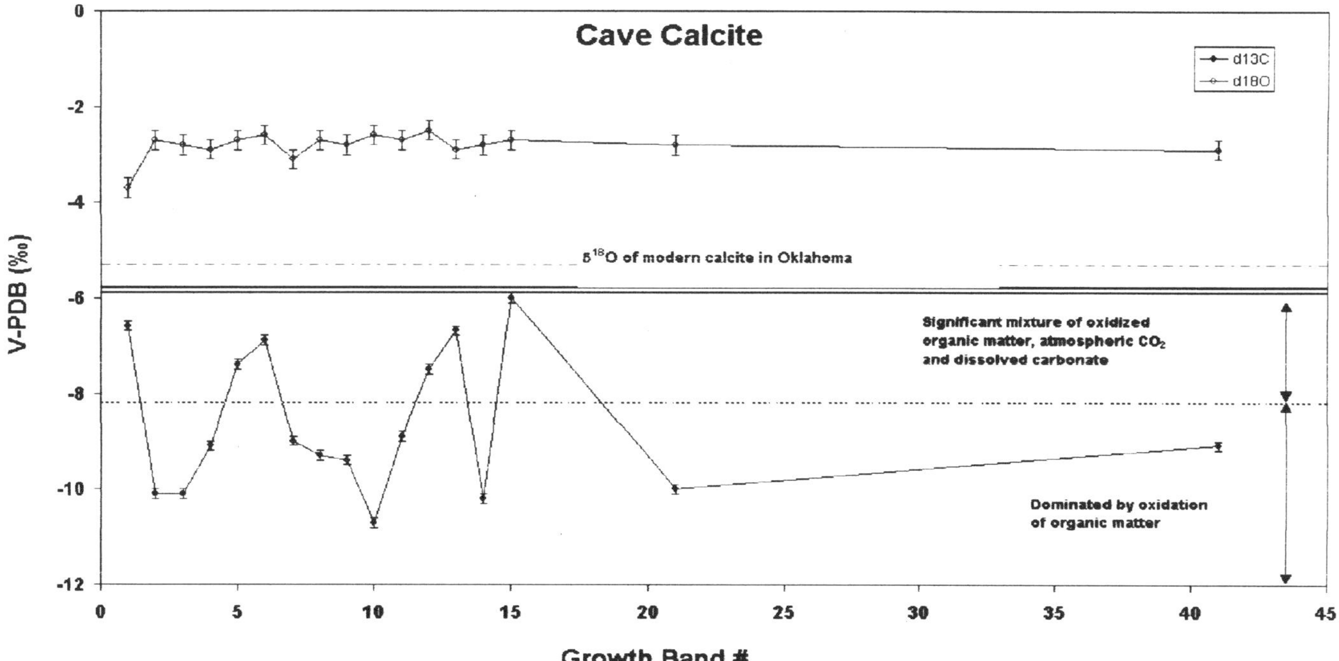

Figure 5. Graph of the measured calcite $\delta^{18}O_{PDB}$ (open circles) and $\delta^{13}C_{PDB}$ (closed circles) values versus the relative position within the sparry calcite sample (see Fig. 3) taken from cave-fill deposit. Oxygen-isotope values are relatively invariant and ~2‰ more positive than the expected $\delta^{18}O_{PDB}$ value for calcites that form in equilibrium with modern meteoric water in this region (solid line; –5.4‰). Carbon-isotope values range from –10.7‰ to –6.6‰, and there appears to be episodic shifts between more positive and negative $\delta^{13}C_{PDB}$ values through the growth series. See text for details and discussion.

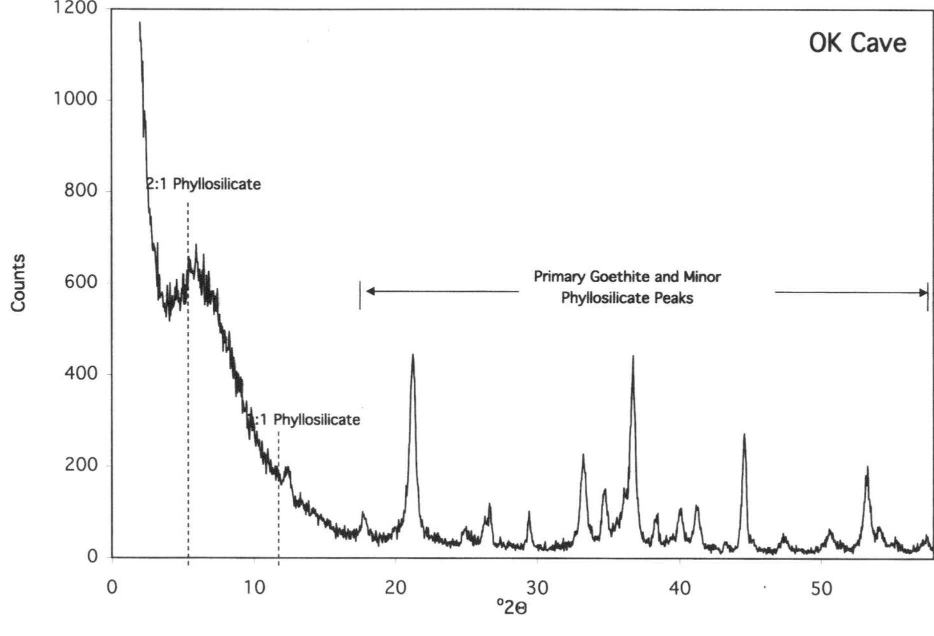

Figure 6. X-ray diffractogram of the OK-Cave sample from the cave-fill deposit. The combination of peaks indicates that this sample dominantly consists of goethite (α-FeOOH) with minor 2:1 and 1:1 phyllosilicate. The d(110) and d(111) peaks of goethite correspond to 3 mol% Al^{3+} substitution for Fe^{3+} in the goethite structure (Schulze, 1984).

TABLE 3. MASS FRACTIONS OF OXYGEN FROM THE VARIOUS OXIDE COMPONENTS IN THE SAMPLES AS WELL AS THE MOLE FRACTION OF OXYGEN ASSOCIATED WITH IRON, $X(O)_{FE}$, AND ALUMINUM $X(O)_{Al}$, FROM GOETHITE FOR EACH SAMPLE

Sample	P_2O_5	K_2O	CaO	MgO	Na_2O	MnO_2	Fe_2O_3	Al_2O_3	SiO_2	TiO_2	H_2O	$X(O)_{Fe}$	$X(O)_{Al}$
OKCave Micro	B.D.	0.0009	0.0007	0.0010	B.D.	0.0004	0.7083	0.0400	0.0276	0.0008	0.2179	0.9444	0.0417
OKCAVE	B.D.	0.0008	0.0001	0.0001	0.0001	0.0003	0.7334	0.0145	0.0127	0.0005	0.2374	.9779	0.0417

Note: B.D.—below detection limit.

TABLE 4. MEASURED OXYGEN AND HYDROGEN ISOTOPE COMPOSITIONS FOR BULK AND RESIDUE FRACTIONS OF THE OKCAVEMICRO AND OKCAVE SAMPLES

Sample	$\delta^{18}O_b$	δD_b	$\delta^{18}O_r$	δD_r	$\delta^{18}O_{go}$	δD_{go}
OKCaveMicro	3.8‰	−124‰	20.3‰	−65‰	3.1 ± 2‰	−125 ± 3‰
OKCAVE	3.0‰	−121‰	21.0‰	−55‰	2.7 ± 2‰	−122 ± 3‰

Note: Also shown are the calculated oxygen and hydrogen isotope compositions for pure end-member goethite in each sample.

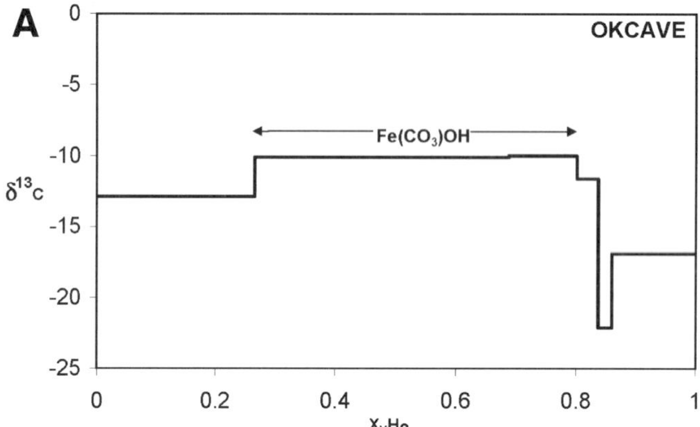

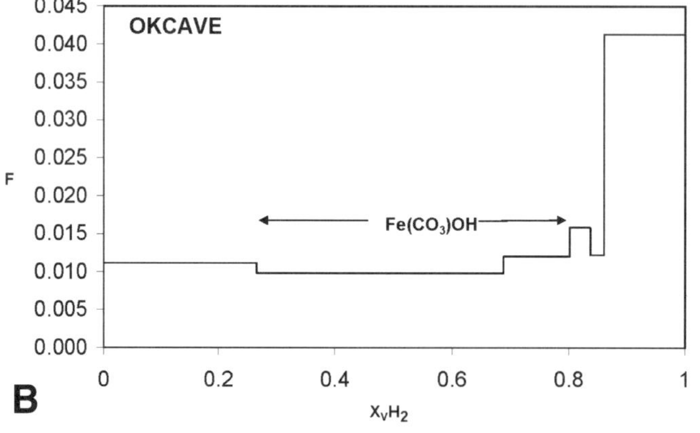

Figure 7. Incremental vacuum dehydration-decarbonation spectra for the $Fe(CO_3)OH$ component in solid solution in OKCave goethite. The graphs present the progress variable $X_v(H_2)$ versus (A) F values (nCO_2/nH_2O) and (B) corresponding $\delta^{13}C$ values of CO_2 evolved during individual increments of sample dehydration. For this sample, there is an apparent plateau in F and $\delta^{13}C$ values over the $X_v(H_2)$ range 0.27–0.80. See text for details and discussion.

(Yapp and Poths, 1991, 1992, 1993; Tabor et al., 2004b). The average plateau $\delta^{13}C$ and F values for OKCaveM are –10.1 ± 0.2‰ and 0.0103 ± 0.0016, respectively (Table 1).

DISCUSSION

Numerous studies have documented a close correspondence between $\alpha^{18}O$ and αD values of goethite, the isotopic composition of ambient waters, and the temperature of goethite crystallization (Yapp, 1987, 1990, 1993, 1997, 2000; Bao and Koch, 1999; Bao et al., 2000; Bird et al., 1992, 1993; Girard et al., 1997, 2000). This relationship may, in turn, be used to reconstruct paleoenvironmental conditions from the $\delta^{18}O$ and δD values of ancient goethite if mineral-water isotope fractionation factors are well known. Partial equilibration and synthesis experiments over the range 25 °C to 145 °C presented in Yapp and Pedley (1985) and Yapp (1987) offer the only available goethite-water hydrogen-isotope fractionation factor [$^{D}\alpha$ = (D/H)goethite/(D/H)water]. Within analytical uncertainty, goethite-water hydrogen-isotope fractionation is:

$$^{D}\alpha = 0.905 \pm 0.004, \qquad (3)$$

and there appears to be no temperature-dependent hydrogen-isotope fractionation. However, several goethite-water oxygen-isotope fractionation factors [$^{18}\alpha_g$ = ($^{18}O/^{16}O$)goethite/($^{18}O/^{16}O$)water] have been proposed (Yapp, 1990; Müller, 1995; Zheng, 1998; Bao and Koch, 1999). Yapp (2000, 2001) concluded that the goethite-water oxygen-isotope fractionation factor presented in Yapp (1990) appears to provide the closest approximation to goethite-water oxygen-isotope fractionation observed in most natural environments to date. This is illustrated by the plot of $^{D}\alpha$ versus $^{18}\alpha_g$ in Figure 8. There are two pairs of goethite isotherms depicted in Figure 8. One pair was calculated using the goethite-water oxygen-isotope fractionation factor of Yapp (1990) for temperatures of 0 °C and 30 °C. The other pair was calculated using the oxygen-isotope fractionation factors of Bao and Koch (1999):

$$10^3 \ln {}^{18}\alpha_g = (1.63*10^6/T_k^2) - 12.3 \text{ (Yapp, 1990)}, \qquad (4)$$

$$10^3 \ln {}^{18}\alpha_g = (1.97*10^3/T_k) - 8.004 \text{ (Bao and Koch, 1999)}, \qquad (5)$$

where T_k is temperature in degrees Kelvin. Both pairs of isotherms in Figure 8 were calculated with the assumption that the goethite formed in the presence of meteoric water as described by the equation of Craig (1961). This global meteoric water line is shown in Figure 8 for reference. The freezing point of pure H_2O at one atm (0 °C) was chosen as the lower temperature limit (cf., Yapp, 2000). The higher temperature of 30 °C was selected because it appears to represent an approximate upper limit to modern mean annual Earth surface air temperatures (e.g., Rozanski et al., 1993). The δD and $\delta^{18}O$ values measured for the OKCave goethite samples are plotted in Figure 8 and lie within the range of values permitted by the Yapp (1987, 1990) fractionation equations. However, these measured data are well outside the range of

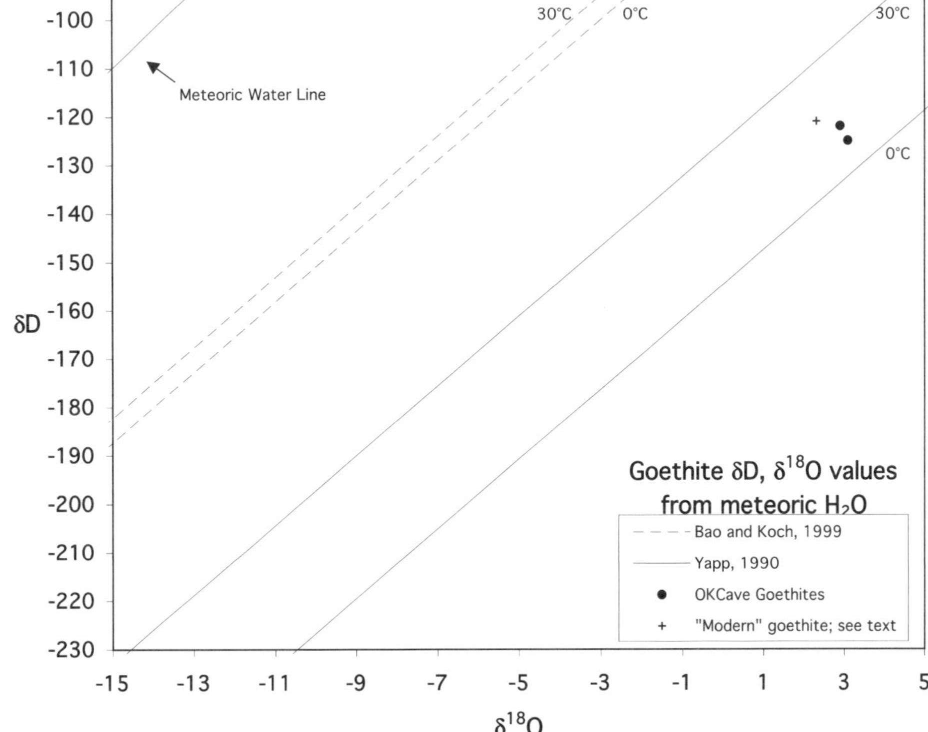

Figure 8. Plot of δD against $\delta^{18}O$ for waters of the meteoric waters (MWL) of Craig (1961). Also, "goethite lines" calculated using the goethite-water hydrogen-isotope fractionation equation of Yapp (1987) and two different goethite-water oxygen-isotope fractionation equations: one proposed by Bao and Koch (1999), the other by Yapp (1990). Also shown are the measured hydrogen- and oxygen-isotope values for OKCave and OKCaveM goethite samples (filled black circles) and a hypothetical "modern" goethite in equilibrium with local meteoric water at mean annual surface air temperature in Oklahoma (cross). The hypothetical goethite δD and $\delta^{18}O$ values were calculated using the oxygen- and hydrogen-isotope fractionation factors of Yapp (1987, 1990). See text for discussion.

values represented by the oxygen-isotope fractionation equation of Bao and Koch (1999) in combination with the goethite-water D/H fractionation equation of Yapp (1987). Further discussion of the data in this work will utilize the goethite-water hydrogen- and oxygen-isotope fractionation equations determined in studies by Yapp (1987, 1990).

SIGNIFICANCE OF GOETHITE D/H AND $^{18}O/^{16}O$ RATIOS

Figure 8 presents the measured oxygen- and hydrogen-isotope compositions of the OKCave and OKCaveM goethites in this study (filled circles). Figure 8 depicts the oxygen- and hydrogen-isotope values expected for goethites in isotopic equilibrium with modern local meteoric waters of southern Oklahoma at present mean annual temperatures (crosses in Fig. 8). Modern meteoric water oxygen- ($\delta^{18}O = -4.8‰$) and hydrogen- ($\delta D = -28‰$) isotope values were taken from the data presented in Kendall and Coplen (2001), whereas the value for mean annual surface air temperature (T = 16.3 °C) was taken from local weather stations in nearby Lawton, Oklahoma, as reported in the National Oceanic and Atmospheric Administration database. The oxygen- and hydrogen-isotope compositions calculated for hypothetical modern goethites at this locale are $\delta^{18}O = 2.3‰$ and $\delta D = -121‰$ (Fig. 9). Within analytical uncertainty, hydrogen-isotope compositions of the OKCave and OKCaveM samples are indistinguishable from values expected for modern goethites forming in southern Oklahoma. The goethite-water hydrogen-isotope fractionation is essentially independent of temperature in the range of temperatures encountered in surface environments (equation 3 above; Yapp, 1987). Therefore, OKCave goethite samples appear to have formed in the presence of waters that were not significantly different in their hydrogen- or oxygen-isotope values from modern meteoric water. However, the measured oxygen-isotope compositions of the OKCave goethites ($\delta^{18}O = 2.9‰, 3.1‰$) are more positive than the value expected for modern goethite ($\delta^{18}O = 2.3‰$). Goethite-water oxygen-isotope fractionation is temperature-dependent (Yapp, 1990), suggesting that these goethites may have formed at temperatures that are different from modern surface air temperatures in southern Oklahoma. Yapp (1987, 1993, 2000) proposed that the oxygen- and hydrogen-isotope composition of goethites might serve as a single-mineral paleothermometer. For the current calculation, it is assumed that the goethite formed in isotopic equilibrium with meteoric waters and the isotopic composition of the samples has not changed since their formation.

The systematic relationship between hydrogen and oxygen isotopes in global meteoric waters (Craig, 1961) is:

$$\delta D_w = 8*\delta^{18}O_w + 10, \tag{6}$$

where the subscript "w" represents meteoric water. Measured δD and $\delta^{18}O$ values of a goethite may be used in combination with equations 2, 3, 4, and 6 to calculate the temperature of goethite crystallization:

$$T = \left(\frac{1.63x10^6}{\Delta^{18}O_{g-w} + 12.3}\right)^{\frac{1}{2}} \tag{7a}$$

$$10^3 \ln^{18}\alpha \approx \delta^{18}O_g - \delta^{18}O_w = \Delta^{18}O_{g-w};\ \delta^{18}O_w = \left\{\frac{\left[\frac{1000+\delta D_g}{{}^D\alpha}\right] - 1010}{8}\right\}$$

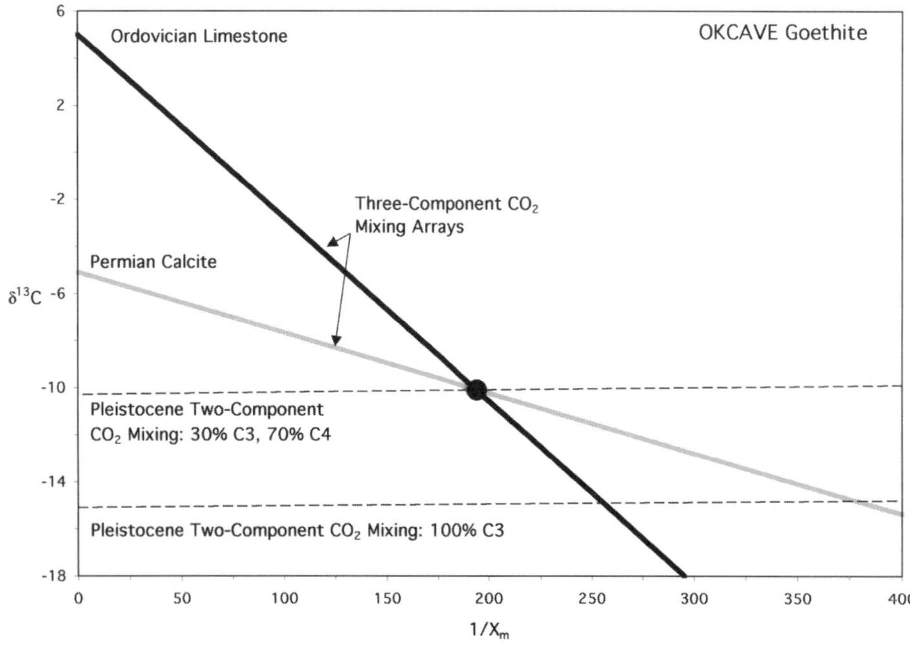

Figure 9. Plot of $\delta^{13}C$ versus $1/X_m$ for the $Fe(CO_3)OH$ component in solid solution in goethite. The measured $1/X_m$ and $\delta^{13}C$ values for the sample OKCaveM (closed circle) lie above expected values for two-component mixing of CO_2 (lower dashed line) in a soil dominated by oxidation of C3 organic matter. Therefore, the $Fe(CO_3)OH$ data probably represent formation of goethite in a system of three-component mixing of CO_2. Possible sources of the third component in this CO_2–mixing system are dissolving Ordovician marine limestone (solid black line) and/or Permian calcite within the cave-fill deposit (solid gray line), or oxidizing organic matter with a significant C4 component (upper dashed line). See text.

where

$$\delta D_g \text{ and } \delta^{18}O_g \quad (7b)$$

are the measured hydrogen- and oxygen-isotope composition, respectively, of goethite, and $^{D}\alpha$ is the stable hydrogen-isotope fractionation factor between goethite and water (0.905). The measured δD_G and $\delta^{18}O_G$ values for OKCave and OKCaveM result in calculated temperatures of 7° ± 3 °C and 11° ± 3 °C, respectively, for goethite crystallization. Considering that shallow subterranean and soil-forming environments typically remain within ±2 °C of mean annual surface temperatures (Buol et al., 1997), it is likely that a calculated temperature of ~9 °C ± 3 °C (average of OKCave and OKCaveM goethites) represents mean annual surface temperatures over the period of goethite crystallization. Note that the calculated temperatures of goethite formation for OKCave and OKCaveM are ~7 °C lower than the current reported mean annual surface air temperature for nearby Lawton, Oklahoma. This suggests that the isotopic composition of the OKCave and OKCaveM goethites records cooler climatic conditions at some time in the geological past. Southern Oklahoma currently resides near its highest latitude for the entire Phanerozoic eon (Scotese and Golonka, 1992). Thus, the low temperatures calculated from the Oklahoma goethites suggest cooler conditions of goethite crystallization either in its current, or more equatorial, geographic position. Lithostratigraphic proxy records around this region indicate a generally warm climate from late Paleozoic through Eocene time (e.g., McGowen et al., 1979; Tabor et al., 2002; Tabor and Moñtanez, 2004; Tabor et al., 2004a). Cenozoic $\delta^{18}O$ records of benthic foraminifera indicate a general Neogene cooling, culminating in globally low temperatures during the Pleistocene glacial maxima (e.g., Zachos et al., 2001). Furthermore, noble gas concentrations of groundwater systems in the Carrizo Aquifer of the northern Gulf coastal plain, Texas, United States (~300 km south of Wichita Mountains), indicate that surface temperatures were 5 °C to 7 °C lower than modern surface temperatures during the last glacial maximum, ~20 ka–10 ka (Stute et al., 1992; Stute and Schlosser, 1993). Notably, cement stratigraphic relationships point to a Pleistocene age of goethite crystallization in these karst-fill deposits of the Wichita Mountains (Donovan et al., 2001). There are no other contemporaneous geochemical proxies of paleoenvironment in this region to compare with this data set. However, palynological, paleobotanical, geomorphic, and paleolake level records around this region do indicate that Pleistocene climate became quite cool and dry (Wells and Stewart, 1987; Fredlund and Jauman, 1987). This cooling was most pronounced during the Pleistocene glacial maximum (Fredlund, 1995). Furthermore, global circulation models suggest that during the Pleistocene glacial maximum the Great Plains region was drier, windier, and 6 °C cooler than current conditions (Kutzbach and Wright, 1985). Therefore, unless the agreement is accidental, the paleotemperature of ~9 °C (i.e., ~7 °C cooler then modern) that was calculated from goethite δD and $\delta^{18}O$ values suggests that these goethites crystallized during one or more glacial stages in the Pleistocene.

SIGNIFICANCE OF CALCITE $\delta^{18}O$ AND $\delta^{13}C$ VALUES

Oxygen Isotopes

The $\delta^{18}O$ value of "fictive" calcite in isotopic equilibrium with modern local meteoric waters at current mean annual temperatures of southern Oklahoma is depicted by the horizontal reference line in Figure 5. This fictive calcite value was calculated using the calcite-water oxygen-isotope fractionation equation of O'Neil et al. (1969). The equation of O'Neil et al. (1969) is

$$10^3 \ln^{18}\alpha_{cc} = (2.78{*}10^6/T_k^{\,2}) - 2.89. \quad (8)$$

The mean measured $\delta^{18}O$ value for the cave calcite is 28.0‰, which is 2.6‰ more positive than the value expected for modern calcite forming in this region (Fig. 5). This indicates that the cave calcites preserve an isotopic record different from the modern. Using a combination of the oxygen-isotope fractionation equations for goethite and calcite (equations 4 and 8) to calculate calcite-goethite mineral pair temperatures yields values of –1 °C and –4 °C. Therefore, the measured oxygen-isotope values of the goethite and calcite do not appear to represent equilibrium crystallization conditions in the presence of the same waters at a common temperature, because the implied temperature would correspond to frozen fresh water in which mineral crystallization and growth of the sort indicated by the calcite and goethite of this study would not occur (cf., Yapp, 2001).

If the calcite did not crystallize in the same environment as goethite, but did form in equilibrium with its own ancient environment, the $\delta^{18}O$ value of the calcite must be considered in terms of two controlling variables of unknown magnitude (temperature and $\delta^{18}O$ of the ambient water, equation 8). As mentioned, there is no geologic evidence to indicate that this region of the world experienced annual temperatures as cool as, or cooler than, Pleistocene glacial stage temperatures at earlier times in the Phanerozoic eon (e.g., Scotese and Golonka, 1992). In addition, cooler mean annual temperatures generally result in more negative isotopic compositions of meteoric precipitation (e.g., Rozanski et al., 1993), which could result in a more negative oxygen-isotope composition of calcite, despite a larger oxygen-isotope fractionation at lower temperatures. Therefore, crystallization of the cave calcite at temperatures significantly cooler than modern is not likely responsible for the measured $\delta^{18}O$ values. As mentioned, the karst-fill sequences of Ordovician limestone have never been deeply buried, and all of the karst-fill calcite cements owe their origin to surface-tied hydrology (Donovan, 1986; Donovan et al., 2001). Furthermore, preservation of Permian vertebrate assemblages and stratigraphic relationships (Olson, 1967) within these karst deposits strongly indicate a Permian age of calcite formation. Based on the oxygen- and hydrogen-isotope compositions of phyllosilicates and goethites from paleosols of the southwestern United States, Permian paleotemperatures of this region ranged from 25 °C to 30 °C (Tabor, 2002). If these temperatures also apply to mineralization of the karst-fill calcites, then the oxygen-isotope composition of the water in equilibrium with these calcites ranged from –0.4‰ to 0.6‰. There are only five modern sites that report such isotopically heavy precipitation (Rozanski et al., 1993). These sites may serve as modern analogs for the environmental conditions associated with calcite precipitation within this fissure-fill deposit. Four out of these five modern sites occur at low latitudes with relatively low mean annual precipitation (<500 mm/yr) and mean annual temperatures ranging from 25.5 °C to 29 °C. These modern analogs correspond well with the paleoequatorial position, calculated paleotemperatures, and inferred paleoprecipitation for this region during early Permian time (Tabor, 2002).

In addition, although cave environments typically have a relative humidity near 100% (Gascoyne, 1992), it is possible that this calcite formed from cave waters that were isotopically modified by evaporation. Furthermore, the effect of diagenetic modification cannot be ignored, in spite of the shallow and low temperature burial history associated with these deposits. Future studies of the hydrogen-isotope composition of fluid inclusion waters in these calcites may help to resolve these issues.

Carbon Isotopes

The cave-calcite sample shows large shifts in $\delta^{13}C$ values through the growth series (Fig. 5) that may be related to temporal changes in mixing of different carbon reservoirs during crystallization (cf., Gascoyne, 1992). In general, there are three primary carbon reservoirs in terrestrial environments that may contribute to the $\delta^{13}C$ composition of newly formed calcite: (1) oxidized biological carbon characterized by relatively negative $\delta^{13}C$ values, (2) inorganic marine and terrestrial carbonate characterized by relatively positive $\delta^{13}C$ values, and (3) atmospheric CO_2 with $\delta^{13}C$ values between biological and inorganic carbonate (e.g., Hoefs, 1997).

There is currently no evidence to suggest that any photosynthetic pathway other than C3 was utilized by terrestrial flora prior to Miocene time (e.g., Cerling, 1991). Biological carbon resulting from C3 photosynthesis has $\delta^{13}C$ values that generally range from ~–29‰ to –23‰ (Cerling and Quade, 1993). The corresponding range of $\delta^{13}C$ values of CO_2 derived from oxidation of C3-biological carbon in soils will range from ~–24.6‰ to –18.6‰, after a 4.4‰ diffusive enrichment (Cerling et al., 1991). At 25 °C, the isotopic composition of calcite will be 10.4‰ more positive than coexisting CO_2 (Bottinga, 1968). This results in a possible range of calcite $\delta^{13}C$ values from –14.2‰ to –8.2‰ in a system dominated by CO_2 gas derived from oxidation of biological material.

The $\delta^{13}C$ value of the ancient atmosphere may be estimated from the carbon-isotope composition of contemporaneous marine calcite (Cerling, 1991). This approach has been used to estimate a range of Permian atmospheric $\delta^{13}C$ values from –4.9‰ to –4.0‰ (Ekart et al., 1999). Thus, in a freshwater system in equilibrium with atmospheric CO_2, $\delta^{13}C$ values of calcite could range from +5.5‰ to +6.4‰.

There is no isotopic fractionation associated with congruent dissolution of carbonate (Wigley et al., 1978). Thus the $\delta^{13}C$ value of HCO_3^- (at pH ~8) derived from dissolution of the host Ordovician limestone in this karst system was likely near the limestone value of +1.5‰. However, equilibrium reprecipitation of calcite at 25 °C from this bicarbonate solution would be 0.9‰ more positive than the dissolved component (Mook et al., 1974), resulting in a $\delta^{13}C$ of +2.4‰ for a newly formed calcite in the fissure-fill deposit derived solely from dissolution of Ordovician carbonate.

Based on the preceding discussion, cave calcite $\delta^{13}C$ values ≤–8.2‰ likely record a system dominated by high, but variable, proportions of CO_2 from oxidation of biological carbon, whereas $\delta^{13}C$ values >–8.2‰ likely record a system with a significant addition of CO_2 from atmospheric CO_2 and/or dissolved limestone (Fig. 5). The $\delta^{13}C$ variation of the cave calcite sample is typical of modern subterranean environments characterized by seasonal changes in rainfall and biological productivity (i.e., in xeric or monsoonal climates; Gascoyne, 1992 and examples therein). Although the exact mechanism for variable $\delta^{13}C$ values in the calcite crystal is not known, changes in the $\delta^{13}C$ value may be related to episodic changes in productivity of the Permian soil mantle above the karst system as a result of seasonal climate variability and changing CO_2 concentrations in the cave atmosphere. This inferred climate system is consistent with paleoclimate reconstructions based on the morphological character of Permian-age paleosols that formed in this region (Tabor and Montañez, 2002).

$\delta^{13}C$ VALUES OF $Fe(CO_3)OH$ IN GOETHITE AND CO_2 MIXING MODELS

Yapp and Poths (1992) presented the following Henry's Law expression for the ferric carbonate component in goethite:

$$\log_{10}P_{CO_2} = \log_{10}X + 6.04 - 1570/T(°K). \quad (9)$$

The X_m value of ferric carbonate in OKCaveM goethite is 0.0052 (Table 1). If goethite in the OKCaveM precipitated at ~9 °C, the ambient partial pressure of CO_2 in the cave would have been ~20,000 ppmV. Such high CO_2 concentrations are typical of cave atmospheres in modern-day cool temperate zones as a result of CO_2 diffusion from high-productivity soils overlying the cave space (Ek and Gewelt, 1985). Therefore, the X_m and $\delta^{13}C$ value of ferric carbonate in the OKCaveM goethite may provide some insight into the productivity of soils overlying the cave atmosphere during Pleistocene time.

On the basis of geological arguments and the measured mole fraction (X_m) and $\delta^{13}C$ values of the $Fe(CO_3)OH$ in goethites of various origins, several studies have concluded that either two-component or three-component CO_2 mixing relations can exist in wet subsurface environments (Hsieh and Yapp, 1999; Yapp, 2001; Tabor et al., 2004b). Two-component mixing assumes that the only two sources contributing to soil CO_2 are CO_2 from the open atmosphere and CO_2 from in situ oxidation of biological carbon. Three-component mixing incorporates an additional source of CO_2 derived from in situ dissolution of preexisting carbonates such as calcite, which is generally enriched in ^{13}C compared to atmospheric CO_2 or CO_2 from oxidation of biological material in the soil (e.g., Hoefs, 1997). These two- and three-component CO_2 mixing models have been used to conceptualize mixing of different CO_2 sources in soils. However, these mixing relations may also apply to CO_2 in caves because these subterranean environments are comprised of CO_2 from the open atmosphere and oxidized organic matter from overlying and juxtaposed soils through diffusive transport (e.g., Gascoyne, 1992). Furthermore, many cave systems will also include some component of CO_2 from the dissolution of host limestones and dolomites as most karst systems form by carbonate dissolution.

Yapp and Poths (1992, 1993) derived the following general equation, expressed in terms of mole fraction (X) and $\delta^{13}C$ value of the $Fe(CO_3)OH$ component in goethite, for mixing of CO_2 derived from the atmosphere and oxidation of organic matter.

$$\delta^{13}C_m = (\delta^{13}C_A - \delta^{13}C_O)X_A/X_m + \delta^{13}C_O, \tag{10a}$$

and

$$\delta^{13}C_O = [\alpha_k\delta^{13}C_B + 1000(\alpha_k - 1)]. \tag{10b}$$

$\delta^{13}C_m$ is the measured $\delta^{13}C$ value of the $Fe(CO_3)OH$ in goethite; $\delta^{13}C_A$ would be the $\delta^{13}C$ value of the $Fe(CO_3)OH$ component if atmospheric CO_2 were the only CO_2 in the soil; X_m is the measured value of X for the $Fe(CO_3)OH$ component in goethite; X_A is the value of X for the $Fe(CO_3)OH$ component if atmospheric CO_2 were the only CO_2 in the soil; $\delta^{13}C_O$ is the value that accounts for the diffusive modification of $\delta^{13}C_B$ (Yapp and Poths, 1993). $\delta^{13}C_B$ is the $\delta^{13}C$ value of the $Fe(CO_3)OH$ if it were in equilibrium with gaseous CO_2 of the same carbon-isotope ratio as the biological carbon being oxidized in the profile (Yapp, 2001). The carbon-isotope fractionation factor between the $Fe(CO_3)OH$ in goethite and CO_2 gas is 1.0025 at 25 °C (Yapp and Poths, 1993). The value α_k is the ratio of the diffusion coefficients (D) of the CO_2 molecules with mass numbers 44 and 45 ($\alpha_k = {}^{44}D/{}^{45}D$). The α_k value in soils is ~1.0044 (Cerling et al., 1991). However, many cave atmospheres appear to circulate through advective flow (Ek and Gewelt, 1985), which could result in α_k values of ~1.0000 for organic matter that is directly oxidized in the cave (i.e., $\delta^{13}C_O = \delta^{13}C_B$; Yapp, 2001). These extreme values of α_k result in two $\delta^{13}C_O$ values for a given value of $\delta^{13}C_B$.

The magnitude of $(\delta^{13}C_A - \delta^{13}C_O)$ in equation 7a is assumed to be about the same at all times in the Phanerozoic for soils dominated by C3 photosynthesis, which is the dominant photosynthetic pathway of plants in the humid climates that favor the formation of goethite (e.g., Hsieh and Yapp, 1999). Yapp and Poths (1996) adopted a value of +16‰ because it represents the approximate difference between preindustrial CO_2 (–6.5‰) and recent C3 continental biota (–27‰), after including a 4.4‰ diffusive ^{13}C enrichment in the biologically derived CO_2 (Cerling et al., 1991). However, as stated earlier, if α_k values for oxidizing organic matter in a cave atmosphere approach ~1.0000, the quantity $(\delta^{13}C_A - \delta^{13}C_O)$ will be 20.4‰. Furthermore, if soil organic matter was characterized by a mixture of 30% C3 and 70% C4 photoynthesizers as suggested by other studies (Fox and Koch, 2004; Koch et al., 2004), then $(\delta^{13}C_A - \delta^{13}C_O)$ would have been a smaller value than for C3 photosynthesis alone. However, an OKCave goethite $Fe(CO_3)OH$ $\delta^{13}C$ value of –10.1‰ constrains $(\delta^{13}C_A - \delta^{13}C_O)$ to be 10.7‰ for a C3-C4 mix with a $\delta^{13}C$ value of –14.7‰ and atmospheric CO_2 concentrations of 180 ppmV. It should be pointed out that error in the selection of a value for $\delta^{13}C_A - \delta^{13}C_O$ has a marginal effect on the concentration of tropospheric CO_2 calculated with equation 10. For example, if the uncertainty in the value of $\delta^{13}C_A - \delta^{13}C_O$ were as large as 4‰, the relative error in the calculated partial pressure of atmospheric CO_2 would be 25%.

Under these proposed conditions, there are two possible end-member two-component mixing equations:

$$\delta^{13}C_m = 10.7{*}X_A/X_m + \delta^{13}C_O, \tag{11a}$$

(all oxidation of C3-C4 mix in the soil),

$$\delta^{13}C_m = 20.4{*}X_A/X_m + \delta^{13}C_B \tag{11b}$$

(all oxidation of pure C3 material in the cave atmosphere).

The $\delta^{13}C$ of naturally occurring modern C3 plants ranges from –29‰ to –23‰ (e.g., Cerling and Quade, 1993). There is no evidence to suggest that Pleistocene atmospheric CO_2 concentrations were much more than preindustrial Holocene values (~300 ppmV) but may be as low as 180 ppmV (e.g., Jouzel et al., 1993; Schlessinger, 1997). As mentioned, the concentration of OKCave ferric carbonate (X_m) is 0.0052, and this sample crystallized at ~9 °C. Therefore, with the values assumed for $\delta^{13}C_B$, α_k, and X_m, equations 11a and 11b indicate that the $\delta^{13}C_m$ value of Pleistocene ferric carbonate in the cave goethite would be no more positive than –14.8‰ for a two-component CO_2 mixing system characterized by oxidation of organic matter derived from C3 photosynthesis (Fig. 9). Therefore, the ferric carbonate $\delta^{13}C_m$ value of –10.1‰ for OKCave indicates this mineral either formed (1) in the presence of oxidizing organic matter derived from mixed C3 and C4 photosynthesizers, (2) in a three-component CO_2 mixing system, regardless of the α_k and the $\delta^{13}C$ value of oxidizing organic matter at the time of goethite crystallization, or (3) a combination of (1) and (2).

On the basis of Pleistocene-age herbivore teeth $\delta^{13}C$ values from northern Texas, United States, Fox and Koch (2004) and Koch et al. (2004) assert that Pleistocene grasslands of the southern Great Plains were occupied by a mixture of ~30% C3 and ~70% C4 grasses. C4 grass $\delta^{13}C$ values typically range from about –9‰ to –15‰ (Teeri and Stowe, 1976). However, considering that OKCave goethite $Fe(CO_3)OH$ has a $\delta^{13}C$ value of –10.1‰ and formed in the presence of Pleistocene atmospheric CO_2 concentrations ≤300 ppmV, and assuming α_K was 1.0044 (i.e., oxidation of organic matter in the soil), the $\delta^{13}C$ value of oxidizing organic matter was probably no greater than –17.2‰ at the time of goethite crystallization (i.e., $\delta^{13}C_O$ = –10.3‰). If the ratio of C3 to C4 plants determined in Koch et al. (2004) is applicable to the OKCave deposit and C3 plants had a relatively heavy $\delta^{13}C$ value (–23‰), then C4 plant $\delta^{13}C$ values would have been about –14.7‰ in the vicinity of this deposit in order to facilitate two component soil CO_2 mixing, yielding a $\delta^{13}C_O$ value of –10.3‰.

The goethite layers in this karst system are situated between Ordovician marine limestone and sparry calcites. The latter likely formed in the phreatic zone. A probable third CO_2 component in OKCave ferric carbonate could have been derived from dissolution of one or both of these potential sources. Fe-sulfides are occluded within some of the sparry cave-calcite deposits (Fig. 4). This suggests that pore waters were sufficiently anoxic for Eh conditions to favor precipitation of Fe^{2+} minerals while also having a pH (~8.3) that favored precipitation of carbonate. However, if these ferrous minerals were subsequently exposed to oxygenated waters and transformed to ferric minerals such as goethite, pore waters within the vicinity of this reaction boundary would become quite acidic and promote carbonate dissolution:

$$4FeS_2 + 15O_2 + 10H_2O = 4FeOOH + 8SO_4^{2-} + 16H^+, \qquad (12a)$$

and

$$8CaCO_3 + 16H^+ = 8Ca^{2+} + 8H_2O + 8CO_2(aq), \qquad (12b)$$

$$Fe^{3+} + 2H_2O + CO_2(aq) = Fe(CO_3)OH + 3H^+. \qquad (12c)$$

Yapp (2001, 2002) presented the following three-component CO_2 mixing equation for the ferric carbonate component in goethite that forms at low pH:

$$\delta^{13}C = [X_A(\delta^{13}C_A - \delta^{13}C_O) + X_S(\delta^{13}C_O - \delta^{13}C_{CC})][1/X_m] + \delta^{13}C_{CC}. \qquad (13)$$

X_A, $\delta^{13}C_A$, $\delta^{13}C_O$, and X_m are the same as in equation 1. X_S is the mole fraction of $Fe(CO_3)OH$ in goethite if it were in equilibrium only with the CO_2 gas in a two-component soil CO_2 mixture (i.e., oxidized biological material and atmospheric CO_2). $\delta^{13}C_{CC}$ is the calculated value of the $Fe(CO_3)OH$ if it were in equilibrium with aqueous CO_2 solely derived from dissolving calcite. The value of $\delta^{13}C_{CC}$ in ferric carbonate will be ~3.5‰ more positive than the $\delta^{13}C$ value of aqueous CO_2 (Yapp, 2001). The potential sources of dissolving carbonate in the presence of OKCave goethite precipitation are Ordovician marine limestone ($\delta^{13}C$ value = +1.5‰) and sparry cave calcite ($\delta^{13}C$ value = –8.6‰), corresponding to $\delta^{13}C_{CC}$ values of 5.0‰ and –5.1‰, respectively (Fig. 9). With low pH three-component mixing involving this range of $\delta^{13}C_{cc}$ values and with the hypothetical two-component mixing lines (equations 11a and 11b), it is possible to calculate values for X_S (equation 13), which provides limits upon the CO_2 concentrations in "upstream" two-component systems representing mixing of CO_2 from Earth's atmosphere and CO_2 from oxidation of organic matter at the time of goethite crystallization. With certain assumed values for $\delta^{13}C_A$, $\delta^{13}C_{cc}$ $\delta^{13}C_O$, and atmospheric P_{CO_2}, CO_2 concentrations in the soils in the vicinity of OKCave goethite would have been between 8,000 ppmV and 16,000 ppmV. This calculated range of soil CO_2 values reflects, in part, variable contributions of CO_2 from assumed endmembers such as dissolving Ordovician- and Permian-age calcite, oxidizing of plant organic matter derived solely from C3 organic matter (i.e., $\delta^{13}C_O$ = –23; Fig. 9) or oxidizing mixed C3:C4 organic matter (i.e., $\delta^{13}C_O$ = –17.2‰; Fig. 9) and α_k values from 1.0044. Such two-component CO_2 concentrations are similar to those found in modern grassland soils of the southern Great Plains (e.g., Wood and Petraitis, 1984). Notably, higher proportions of C4 organic matter result in the higher estimates of CO_2 concentrations contributed from oxidation of organic matter during OKCave goethite crystallization. The isotopic results of the OKCave and OKCaveM goethites are intriguing, as these data imply a relatively high-productivity soil in the southern plains at a time of significantly cooler Pleistocene climate.

CONCLUSIONS

δD, $\delta^{18}O$, and $\delta^{13}C$ values of secondary calcites and goethites preserved within a karst-fill deposit in the Wichita Mountains, south-central Oklahoma, are suggestive of formation under two distinctly different environmental conditions. Prior studies have suggested that calcites in the fissure fill formed during early Permian time, whereas the goethites formed from dissolution and oxidation of Fe-sulfides and reprecipitation of iron oxyhydroxides during Pleistocene time. Calcite from a cave-fill deposit has relatively invariant oxygen-isotope compositions suggestive of formation in a warm and dry, seasonal, coastal environment. Furthermore, large, episodic shifts in the carbon-isotope composition of calcite in this deposit may record seasonal changes in biological productivity of the overlying soil mantle in a Permian monsoonal or xeric climate. These conclusions are consistent with other Permian paleoclimate reconstructions inferred from lithologic and isotopic proxy data in this region.

Goethites from the cave-fill deposit are not in isotopic equilibrium with modern waters of south-central Oklahoma nor with coexisting calcites. The combined $\delta^{18}O$ and δD values of the goethites indicate that crystallization occurred at ~9 °C, which is ~7 °C cooler than modern temperatures in this region. This lower temperature is in reasonable accord with a late Pleistocene glacial stage temperature ~6 °C cooler than modern, which was determined from groundwater noble gas concentrations in the United States Gulf Coastal Plain (Stute et al., 1992; Stute and Schlosser, 1993), and also with a glacial maximum temperature that was ~6 °C cooler in the Pleistocene in the southern Great Plains, as predicted from general circulation models (Kutzbach and Wright, 1985).

The carbon-isotope composition of the $Fe(CO_3)OH$ component in solid solution in goethite was probably acquired in a system in which three distinct CO_2 components were mixing. The combination of CO_2 derived from atmospheric CO_2 and oxidation of biological material likely totaled no less than 8,000 ppmV and no greater than 16,000 ppmV in this system.

ACKNOWLEDGMENTS

This research was funded in part by NSF grant EAR-0106257 to C.J. Yapp. We thank Paul Koch and Paul Schroeder for helpful reviews.

REFERENCES CITED

Bao, H., and Koch, P.L., 1999, Oxygen isotope fractionation in ferric oxide-water systems: Low temperature synthesis: Geochimica et Cosmochimica Acta, v. 63, p. 599–613, doi: 10.1016/S0016-7037(99)00005-8.

Bao, H., Koch, P.L., and Thiemens, M.H., 2000, Oxygen isotope composition of ferric oxides from recent soil, hydrologic, and marine environments: Geochimica et Cosmochimica Acta, v. 64, p. 2221–2231, doi: 10.1016/S0016-7037(00)00351-3.

Bigeleisen, J., Perlman, M.L., and Prosser, H.C., 1952, Conversion of hydrogenic materials to hydrogen for isotopic analysis: Analytical Chemistry, v. 24, p. 1356–1357, doi: 10.1021/ac60068a025.

Bird, M.I., and Chivas, A.R., 1988, Oxygen isotopic dating of the Australian Regolith: Nature, v. 331, p. 513–516, doi: 10.1038/331513a0.

Bird, M.I., Longstaffe, F., Fyfe, W.S., and Bildgen, P., 1992, Oxygen isotope systematics in a multiphase weathering system in Haiti: Geochimica et Cosmochimica Acta, v. 56, p. 2831–2838, doi: 10.1016/0016-7037(92)90362-M.

Bird, M.I., Longstaffe, F., Fyfe, W.S., Kronberg, B.I., and Kisida, A., 1993, An oxygen isotope study of weathering in the eastern Amazon Basin, Brazil, *in* Swart, P.K., Lohman, K.C., McKenzie, J., and Savin, S., eds., Climate Change in Continental Isotopic Records: Washington, D.C., American Geophysical Union, Geophysical Monograph 78, p. 295–307.

Bottinga, Y., 1968, Calculation of fractionation factors for carbon and oxygen in the system calcite–carbon dioxide–water: Journal of Physical Chemistry, v. 72, p. 800–808, doi: 10.1021/j100849a008.

Buol, S.W., Hole, F.D., McCracken, R.J., and Southard, R.J., 1997, Soil Genesis and Classification: Ames, Iowa, Iowa State University Press, 527 p.

Cerling, T.E., 1991, Carbon dioxide in the atmosphere: Evidence from Cenozoic and Mesozoic paleosols: American Journal of Science, v. 291, p. 377–400.

Cerling, T.E., and Quade, J., 1993, Stable carbon and oxygen isotopes in soil carbonates, *in* Swart, P.K., Lohman, K.C., McKenzie, J., and Savin, S., eds., Climate Change in Continental Isotopic Records: Washington, D.C., American Geophysical Union, Geophysical Monograph 78, p. 217–231.

Cerling, T.E., Solomon, D.K., Quade, J., and Bowman, J.R., 1991, On the isotopic composition of carbon in soil carbon dioxide: Geochimica et Cosmochimica Acta, v. 55, p. 3403–3405, doi: 10.1016/0016-7037(91)90498-T.

Clayton, R.N., and Mayeda, T.K., 1963, The use of bromine pentafluoride in the extraction of oxygen from oxides and silicates for isotopic analysis: Geochimica et Cosmochimica Acta, v. 27, p. 43–52, doi: 10.1016/0016-7037(63)90071-1.

Craig, H., 1957, Isotopic standards for carbon and oxygen and correction factors for mass-spectrometric analysis of carbon dioxide: Geochimica et Cosmochimica Acta, v. 12, p. 133–149, doi: 10.1016/0016-7037(57)90024-8.

Craig, H., 1961, Isotopic variations in meteoric waters, Science, v. 133, p. 1702–1703.

Delgado, A., and Reyes, E., 1996, Oxygen and hydrogen isotope compositions in clay minerals; a potential single-mineral paleothermometer: Geochimica et Cosmochimica Acta, v. 60, p. 4285–4289, doi: 10.1016/S0016-7037(96)00260-8.

Donovan, R.N., 1986, The geology of the Slick Hills, *in* Donovan, R.N., ed., The Slick Hills of southwestern Oklahoma—Fragments of an aulacogen?: Oklahoma Geological Survey Guidebook 24, p. 1–12.

Donovan, R.N., Collins, K., and Bridges, S., 2001, Permian Sedimentation and Diagenesis on the Northern Margin of the Wichita Uplift: Oklahoma Geological Survey Circular 104, p. 171–184.

Ek, C., and Gewelt, M., 1985, Carbon dioxide in cave atmospheres: New Results in Belgium and comparison with some other countries: Earth Surface Processes and Landforms, v. 10, p. 173–187.

Ekart, D.D., Cerling, T.E., Montanez, I.P., and Tabor, N.J., 1999, A 400 million year carbon isotope record of pedogenic carbonate: Implications for paleoatmospheric carbon dioxide: American Journal of Science, v. 299, p. 805–827.

Fox, D.L., and Koch, P.L., 2004, Carbon and oxygen isotopic variability in Neogene paleosol carbonates: Constraints on the evolution of the C4-grasslands of the Great Plains, USA: Palaeogeography, Palaeoclimatology, Palaeoecology, v. 207, p. 305–330, doi: 10.1016/S0031-0182(04)00045-8.

Fredlund, G.G., 1995, Late Quaternary pollen record from Cheyenne Bottoms, Kansas: Quaternary Research, v. 43, p. 67–79, doi: 10.1006/qres.1995.1007.

Fredlund, G.G., and Jauman, P.J., 1987, Late Quaternary palynological and paleobotanical records from the central Great Plains, *in* W.C. Johnson, ed., Quaternary Environment of Kansas, Kansas Geological Survey Guidebook series 5, p. 167–178.

Gascoyne, M., 1992, Palaeoclimate determination from cave calcite deposits: Quaternary Science Reviews, v. 11, p. 609–632, doi: 10.1016/0277-3791(92)90074-I.

Gilg, H.A., 2000, D-H evidence for the timing of kaolinitization in NE Bavaria, Germany: Chemical Geology, v. 170, p. 5–18, doi: 10.1016/S0009-2541(99)00239-9.

Girard, J.P., Razandranoroso, D., and Freyssinet, P., 1997, Laser oxygen isotope analysis of weathering goethite from the lateritic profile of Yaou, French Guiana: Paleoweathering and paleoclimatic implications: Applied Geochemistry, v. 12, p. 163–174, doi: 10.1016/S0883-2927(96)00062-5.

Girard, J.P., Feyssinet, P., and Chazot, G., 2000, Unraveling climatic change from intra-profile variation in oxygen and hydrogen isotopic composition of goethite and kaolinite in laterites: An integrated study from Yaou, French Guiana: Geochimica et Cosmochimica Acta, v. 64, p. 409–426, doi: 10.1016/S0016-7037(99)00299-9.

Gonfiantini, R., 1984, Advisory group meeting on stable isotope reference samples for geochemical and hydrological investigations: Rep Director General International Atomic Energy Association Vienna.

Hoefs, J., 1997, Stable Isotope Geochemistry: Berlin, Springer Verlag, 200 p.

Hsieh, J.C.C., and Yapp, C.J., 1999, Stable carbon isotope budget of CO_2 in a wet, modern soil as inferred from $Fe(CO_3)OH$ in pedogenic goethite: Possible role of calcite dissolution: Geochimica et Cosmochimica Acta, v. 63, p. 767–783, doi: 10.1016/S0016-7037(99)00062-9.

Jackson, M.L., 1979, Soil Chemical Analysis—Advanced Course: Published by the author, Madison, Wisconsin.

Jouzel, J., Barkov, N.I., Barnola, J.M., Bender, M., Chappellaz, J., Genthon, C., Kotlyakov, V.M., Lipenkov, V., Loerius, C., Petit, J.R., Raynaoud, D., Raisbeck, G., Ritz, C., Sowers, T., Stievenard, M., Yiou, F., and Yiou, P., 1993, Extending the Vostok ice-core record of palaeoclimate to the penultimate glacial period: Nature, v. 364, p. 407–412.

Kendall, C., and Coplen, T.B., 2001, Distribution of oxygen-18 and deuterium in river water across the United States: Water Quality of large U.S. rivers; results from the U.S. Geological Survey's National Stream Quality Accounting Network: Hydrological Processes, v. 15, p. 1363–1393, doi: 10.1002/hyp.217.

Koch, P.L., Diffenbaugh, N.S., and Hoppe, K.A., 2004, The effects of late Quaternary climate and pCO_2 change on C4 plant abundance in the south-central United States: Palaeogeography, Palaeoclimatology, Palaeoecology, v. 207, p. 331–358, doi: 10.1016/S0031-0182(04)00046-X.

Kutzbach, J.E., and Wright, H.E., Jr., 1985, Simulation of the climate of 18,000 years BP: Results for the North American/North Atlantic/European sector and comparison with the geological record of North America: Quaternary Science Reviews, v. 4, p. 147–187, doi: 10.1016/0277-3791(85)90024-1.

Lawrence, J.R., and Rashkes-Meaux, J., 1993, The stable isotopic composition of ancient kaolinites of North America, *in* Swart, P.K., Lohman, K.C., McKenzie, J., and Savin, S., eds., Climate Change in Continental Isotopic Records: Washington, D.C., American Geophysical Union, Geophysical Monograph 78, p. 249–261.

Ludvigson, G.A., 1998, Meteoric sphaerosiderite lines and their use for paleohydrology and paleoclimatology: Geology, v. 26, p. 1039–1042, doi: 10.1130/0091-7613(1998)026<1039:MSLATU>2.3.CO;2.

McGowen, J.H., Granata, G.E., and Seni, S.J., 1979, Depositional framework of the Lower Dockum Group (Triassic): Austin, Texas, Report of Investigations 97, Bureau of Economic Geology.

Mook, W.G., Bommerson, J.C., and Staverman, W.H., 1974, Carbon isotope fractionation between dissolved bicarbonate and gaseous carbon dioxide: Earth and Planetary Science Letters, v. 22, p. 169–176, doi: 10.1016/0012-821X(74)90078-8.

Moore, D.M., and Reynolds, R.C., 1997, X-ray diffraction and the identification and analysis of clay minerals: New York, Oxford University Press, 378 p.

Müller, J., 1995, Oxygen isotopes in iron (III) oxides: A new preparation line; mineral-water fractionation factors and paleoenvironmental considerations: Isotopes in Environment and Health Studies, v. 31, p. 301–302.

O'Neil, J.R., Clayton, R.N., and Mayeda, T.K., 1969, Oxygen isotope fractionation in divalent metal carbonates: The Journal of Chemical Physics, v. 51, p. 5547–5558, doi: 10.1063/1.1671982.

Olson, E.C., 1967, Early Permian Vertebrates: University of Oklahoma Press, Oklahoma Geological Survey circular 74, 111 p.

Poage, M.A., and Chamberlain, C.P., 2001, Empirical relationships between elevation and the stable isotope composition of precipitation and surface waters: Considerations for studies of paleoelevation change: American Journal of Science, v. 301, p. 1–15.

Rozanski, K., Araguas-Araguas, L., and Gonfiantini, R., 1993, Isotopic patterns in modern global precipitation, *in* Swart, P.K., Lohman, K.C., McKenzie, J., and Savin, S., eds., Climate Change in Continental Isotopic Records: Washington, D.C., American Geophysical Union, Geophysical Monograph 78, p. 1–36.

Savin, S.M., and Epstein, S., 1970, The oxygen and hydrogen isotope geochemistry of clay minerals: Geochimica et Cosmochimica Acta, v. 34, p. 25–42, doi: 10.1016/0016-7037(70)90149-3.

Schlessinger, W.H., 1997, Biogeochemistry: An Analysis of Global Change: San Diego, Academic Press, 588 p.

Schulze, D.G., 1984, The influence of aluminum on iron oxides. VIII. Unit cell dimensions of Al-substituted goethites and estimation of aluminum from them: Clays and Clay Minerals, v. 32, p. 36–44.

Scotese, C.R., and Golonka, J., 1992, Paleogeographic Atlas: Arlington, Texas, Paleomap Project, University of Texas-Arlington.

Stute, M., and Schlosser, P., 1993, Principles and applications of the noble gas paleothermometer, *in* Swart, P.K., Lohman, K.C., McKenzie, J., and Savin, S., eds., Climate Change in Continental Isotopic Records: Washington, D.C., American Geophysical Union, Geophysical Monograph 78, p. 89–100.

Stute, M., Schlosser, P., Clark, J.F., and Broecker, W.S., 1992, Paleotemperatures in the southwestern United States derived from noble gas measurements in groundwater: Science, v. 256, p. 1000–1003.

Tabor, N.J., 2002, Paleoclimate isotopic proxies derived from Paleozoic, Mesozoic, Cenozoic and modern soils [Ph. D. dissert.]: Davis, California, University of California, 213 p.

Tabor, N.J., and Montañez, I.P., 2002, Shifts in late Paleozoic atmospheric circulation over western equatorial Pangea: Insights from pedogenic mineral $\delta^{18}O$ compositions: Geology, v. 30, p. 1127–1130, doi: 10.1130/0091-7613(2002)030<1127:SILPAC>2.0.CO;2.

Tabor, N.J., and Montañez, I.P., 2004, Morphology and distribution of fossil soils in the Permo-Pennsylvanian Wichita and Bowie Groups, north-central Texas, USA; implications for western equatorial Pangean palaeoclimate during icehouse-greenhouse transition, Sedimentology, v. 51, p. 851–884.

Tabor, N.J., Montañez, I.P., and Southard, R.J., 2002, Mineralogical and stable isotopic analysis of pedogenic proxies in Permo-Pennsylvanian paleosols: Implications for paleoclimate and paleoatmospheric circulation: Geochimica et Cosmochimica Acta, v. 66, p. 3093–3107.

Tabor, N.J., Montañez, I.P., Zierenberg, R., and Currie, B.S., 2004a, Mineralogical and geochemical evolution of a basalt-hosted fossil soil (Late Triassic, Ischigualasto Formation, Northwest Argentina): Potential for paleoenvironmental reconstruction: Geological Society of America Bulletin, v. 116, p. 1280–1293, doi: 10.1130/B25222.1.

Tabor, N.J., Yapp, C.J., and Montañez, I.P., 2004b, Goethite, calcite, and organic matter from Permian and Triassic soils: Carbon isotopes and CO_2 concentrations: Geochimica et Cosmochimica Acta, v. 68, p. 1503–1517, doi: 10.1016/S0016-7037(03)00497-6.

Teeri, J., and Stowe, L.G., 1976, Climatic patterns and the distribution of C4 grasses in North America: Oecologia, v. 23, p. 1–12.

Wells, P.V., and Stewart, J.D., 1987, Cordilleran-boreal taiga and fauna on the central Great Plains of North America, 14,000–18,000 years ago: American Midland Naturalist, v. 188, p. 94–106.

Wigley, T.M.L., Plummer, L.N., and Pearson, F.J., Jr., 1978, Mass transfer and carbon isotope evolution in natural water systems: Geochimica et Cosmochimica Acta, v. 42, p. 1117–1139, doi: 10.1016/0016-7037(78)90108-4.

Wood, W.W., and Petraitis, M.J., 1984, Origin and distribution of carbon dioxide in the unsaturated zone of the southern high plains of Texas: Water Resources Research, v. 20, p. 1193–1208.

Yapp, C.J., 1987, Oxygen and hydrogen isotope variations among goethites (α-FeOOH) and the determination of paleotemperatures: Geochimica et Cosmochimica Acta, v. 51, p. 355–364, doi: 10.1016/0016-7037(87)90247-X.

Yapp, C.J., 1990, Oxygen isotopes in iron (III) oxides. 1. Mineral-water fractionation factors: Chemical Geology, v. 85, p. 329–335, doi: 10.1016/0009-2541(90)90010-5.

Yapp, C.J., 1993, The stable isotope geochemistry of low temperature Fe (III) and Al "oxides" with implications for continental paleoclimates, *in* Swart, P.K., Lohman, K.C., McKenzie, J., and Savin, S., eds., Climate Change in Continental Isotopic Records: Washington, D.C., American Geophysical Union, Geophysical Monograph 78, p. 285–294.

Yapp, C.J., 1997, An assessment of isotopic equilibrium in goethites from a bog iron deposit and a lateritic regolith: Chemical Geology, v. 135, p. 159–171, doi: 10.1016/S0009-2541(96)00112-X.

Yapp, C.J., 1998, Paleoenvironmental interpretations of oxygen isotope ratios in oolitic ironstones: Geochimica et Cosmochimica Acta, v. 62, p. 2409–2420, doi: 10.1016/S0016-7037(98)00164-1.

Yapp, C.J., 2000, Climatic implications of surface domains in arrays of δD and $\delta^{18}O$ from hydroxyl minerals: Goethite as an example: Geochimica et Cosmochimica Acta, v. 64, p. 2009–2025, doi: 10.1016/S0016-7037(00)00347-1.

Yapp, C.J., 2001, Mixing of CO_2 in surficial environments as recorded by the concentration and $\delta^{13}C$ values of the $Fe(CO_3)OH$ component in goethite: Geochimica et Cosmochimica Acta, v. 65, p. 4115–4130, doi: 10.1016/S0016-7037(01)00698-6.

Yapp, C.J., 2002, Erratum to Crayton J. Yapp (2001), Mixing of CO_2 in surficial environments as recorded by the concentration and $\delta^{13}C$ values of the $Fe(CO_3)OH$ component in goethite: Geochimica et Cosmochimica Acta, v. 66, p. 1497, doi: 10.1016/S0016-7037(01)00889-4

Yapp, C.J., and Pedley, M.D., 1985, Stable hydrogen isotopes in iron oxides-II: D/H variations among natural goethites: Geochimica et Cosmochimica Acta, v. 49, p. 487–495, doi: 10.1016/0016-7037(85)90040-7.

Yapp, C.J., and Poths, H., 1991, $^{13}C/^{13}C$ ratios of the Fe (III) carbonate component in natural goethites, *in* Taylor, H.P., Jr., O'Neil, J.R., and Kaplan, I.R., eds., Stable Isotope geochemistry: A Tribute to Samuel Epstein, Geochemical Society Special Publication 3, p. 257–270.

Yapp, C.J., and Poths, H., 1992, Ancient atmospheric CO_2 pressures inferred from natural goethites: Nature, v. 355, p. 342–344, doi: 10.1038/355342a0.

Yapp, C.J., and Poths, H., 1993, The carbon isotope geochemistry of goethite (α-FeOOH) in ironstone of the Upper Ordovician Neda Formation, Wisconsin, USA: Implications for early Paleozoic continental environments: Geochimica et Cosmochimica Acta, v. 57, p. 2599–2611, doi: 10.1016/0016-7037(93)90420-2.

Yapp, C.J., and Poths, H., 1996, Carbon isotopes in continental weathering environments and variations in ancient atmospheric CO_2 pressure: Earth and Planetary Science Letters, v. 137, p. 71–82.

Zachos, J., Pagani, M., Sloan, L., Thomas, E., and Billups, K., 2001, Trends, rhythms, and aberration in global climate 65 Ma to present: Science, v. 292, p. 686–693, doi: 10.1126/science.1059412.

Zheng, Y.-F., 1998, Oxygen isotope fractionation between hydroxide minerals and water: Physics and Chemistry of Minerals, v. 25, p. 213–221, doi: 10.1007/s002690050105.

Manuscript Accepted by the Society 19 April 2005

Printed in the USA